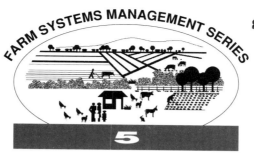

Institutionalization of a farming systems approach to development

Proceedings of
technical discussions
Rome, 15-17 October 1991

FOOD AND AGRICULTURE ORGANIZATION OF THE UNITED NATIONS
Rome, 1992

The designations employed and the presentation of material in this publication do not imply the expression of any opinion whatsoever on the part of the Food and Agriculture Organization of the United Nations concerning the legal status of any country, territory, city or area or of its authorities, or concerning the delimitation of its frontiers or boundaries.

M-61
ISBN 92-5-103250-5

All rights reserved. No part of this publication may be reproduced, stored in a retrieval system, or transmitted in any form or by any means, electronic, mechanical, photocopying or otherwise, without the prior permission of the copyright owner. Applications for such permission, with a statement of the purpose and extent of the reproduction, should be addressed to the Director, Publications Division, Food and Agriculture Organization of the United Nations, Viale delle Terme di Caracalla, 00100 Rome, Italy.

© FAO 1993

FOREWORD

Farming Systems have been on the development agenda now for over a decade, during which time there have been changes in emphasis and fluctuations in its popularity. The main focus of the approach has, however, continued to be technology generation and research.

In FAO, Farming Systems thinking has been promoted in a much wider context than hitherto propagated: Farming Systems Development (FSD), or a Farming Systems approach to Development, is designed to improve such important development areas as Research, Extension, Support Services and Policy analysis; in many cases embedded in Development programmes and projects.

The Farm Management and Production Economics Service of FAO has defined the FSD concept, developed appropriate methodologies and refined them for specific applications. The issue of institutionalization arose as a result of deliberations on i) how to get the approach quickly and effectively applied by all agents in development; and ii) how to change operational procedures and inter- and intra-institutional relations in order to improve the development situation in rural areas.

In pursuit of these topics, Technical Discussions were held in FAO, Rome, from 15-17 October 1991, organized by K.-H. Friedrich, Senior Officer (Farming Systems), and Malcolm Hall, consultant, serving as main resource person. The Technical Discussions were attended by participants with substantial experience in Farming Systems, in search for ways and means to get this approach widely utilized, be it in research, extension, education or administration. The views expressed in the papers are those of the authors.

These Technical Discussions highlighted possible ways forward for Farming Systems. We consider it important to share our findings with other practitioners in development, interested in the subject. We feel that we have made some headway, but also realize that there is still a great deal to be accomplished. In order to advance the subject we have documented the outcome of the Technical Discussions and would appreciate any further constructive contributions from our readers.

Please, send your comments to the following address:

Chief, Farm Management and Production Economics Service,
Agricultural Services Division,
Food and Agriculture Organization of the United Nations,
Via delle Terme di Caracalla
00100 Rome, ITALY

INSTITUTIONALIZATION OF A FARMING SYSTEMS APPROACH TO DEVELOPMENT

TABLE OF CONTENTS

		page
	FOREWORD	
I	**INTRODUCTION**, H. de Haen	1
II	**ISSUES AND OPTIONS IN INSTITUTIONALIZATION OF A FARMING SYSTEMS APPROACH**	9
	Overview and Synthesis	11
	Incorporating the Farming Systems Approach into the Work of Agricultural Development Institutions Hall, M. and K.-H. Friedrich	15
	The Institutionalization of Farming Systems Development, A. Shepherd	29
	Farming Systems Development - Can it be Institutionalized, N.R.C. Ranaweera	59
	Farming Systems Research and Rural Poverty: A Political Economy Perspective on Institutionalization, S.D. Biggs	65
III	**ADOPTING FARMING SYSTEMS APPROACHES IN EDUCATIONAL INSTITUTIONS**	85
	Overview and Synthesis	87
	Farming Systems Development in Academic Training Doppler. W. and M. Maurer	91
	Institutionalization of the Farming Systems Research and Extension Approach in the Agricultural Training System in Tanzania, A.Z. Mattee	107
	Farming System Research: Development of Inter-disciplinary Research at Brawijaya University, Malang, Indonesia, I. Semaoen, A. Liiek and G. Zemmelink	119

IV APPLYING FARMING SYSTEMS CONCEPTS TO TECHNOLOGY DEVELOPMENT — 127

Overview and Synthesis — 129

Strategies for Incorporating Farming Systems Research and Development into Institutional Structures and Organisations, D. Gibbon — 133

Institutionalization of Farming Systems Approach in Agricultural Research and Development set up in Eastern India, V.P. Singh — 161

Research and Development: Farming Systems Research at the Service of Rural Development, Jouve, P. and M.R. Mercoiret — 193

Attempts to Incorporate/Institutionalize an FSR/E Approach into the Research Function of an Agricultural University (VISCA) in the Philippines, Ly Tung — 201

V FARMING SYSTEMS, GOVERNMENTS AND NGO's — 213

Overview and Synthesis — 215

Institutionalization of Farming Systems Development. Are there Lessons from NGO-Government Links?, Farrington J. and A. Bebbington — 219

VI SOME COUNTRY EXPERIENCES WITH FARMING SYSTEMS APPROACHES — 275

Overview and Synthesis — 277

Institutionalization of FSD as Part of Rural Development, B.N. de los Reyes — 281

Institutionalizing Farming Systems Research and Extension in the Philippines, W. Dar — 287

Agro-Ecosystem and Farming Systems Development, their Significance to Agricultural Development in China, Z. Wang — 305

Farming Systems Approach - Kenya Experience, G.G. Mwangi — 315

VII	SUMMARY AND CONCLUSIONS	321
	APPENDIX LIST OF PARTICIPANTS	329

INTRODUCTION

INTRODUCTION

by

H. de Haen[1]

INTRODUCTION

1. THE FARMING SYSTEMS APPROACH

Farming systems concepts have featured prominently in discussions on agricultural development during the past decade. Many useful approaches and techniques have been developed under such headings as "Farming Systems Analysis", "Farming Systems Research", and "Farming Systems Research and Extension". FAO has added its approach to the list, namely "Farming Systems Development (FSD)".

Despite the diversity of "labels", and the fact that the various schools of thought tend to focus on different aspects of the process of agricultural development, all farming systems approaches have a common conceptual core. This could be summarized as follows:

- The farm-household is the basic unit of decision-making and hence analysis in most agricultural development situations. Though systems interactions can be studied at different degrees of aggregation all the way to the national level, the farming systems approach is essentially focused at the grassroots level, upon the rural family.

- Since decisions at household level encompass both production and consumption, interactions between farm work, household tasks and non-farm income earning activities are important. They compete for

---

[1] Assistant Director General, Agriculture Department, FAO

the limited labour and capital resources of the farm family, and all have a bearing on the welfare of rural people.

- The physical, socio-cultural, institutional and economic environment has an important influence upon farm-household decisions and upon their outcomes, but vice versa this environment is also affected by the activities of the farm-household. The interdependence of farm and environment is thus dynamic.

- The rural community is a source of knowledge and the participation of the rural people is essential for the success of any development effort.

- Any improvement measures involve trade-offs between competing household objectives, and these objectives are not necessarily consonant with the sectoral priorities of governments. Alternative innovations and solutions to problems should, therefore, be weighed against their outcomes in terms of both household and social objectives (sectoral and macro).

- Farming systems approaches are holistic and interdisciplinary, as any farming system is affected by an extremely wide range of interactions covering technical, scientific, socio-political and economic factors. Each facet adds a new perspective to the analysis of the rural situation.

The farming systems approach involves the search for order in a world of seeming chaos! The overwhelming complexity of farming systems makes it essential to distinguish patterns, envisage modal situations and simplify approaches through the establishment of broad sequences and indicative checklists. The real world, however, contains variability over time and place, conflicts between groups' and individuals' objectives, competition for resources, spontaneous technological breakthroughs, and constant difficulties of measurement and monitoring.

Farmers have a holistic approach to any problem related to their farms. Their problems need to be approached with a basic understanding of the array of factors that enter into their own analysis and decision-making. A narrower approach concentrating upon a single commodity, or by otherwise applying a partial analysis, would almost certainly overlook key interactions with the rest of the system. These interactions must be understood if sound advice is to be given and the often unspoken rationale of farmers fully comprehended.

FAO's Farming Systems Development (FSD) approach improves upon traditional component and commodity approaches. It starts with an attempt to understand the interactions, and to analyze the major constraints and potentials, within a particular farm-household system. Solutions can be found in any of the activities of the actual farm-household: farming and domestic activities, as well as non-agricultural (on-farm or off-farm) income earning activities. The approach attempts to identify the impact of changes in agricultural policies and marketing structures on farm income and rural family welfare. It also addresses the effect of changes in physical resource endowments on the farm-household. Thus it is relevant to environmental considerations and especially to the sustainability of the natural resource base. In the case of common property resources, solutions might be located at the community level, whereas institutional or policy problems might have to be addressed at higher levels of the systems hierarchy.

2. APPLICATION OF THE FARMING SYSTEMS APPROACH

Variations of this farming systems approach are increasingly being adopted by institutions covering a broad spectrum of agricultural development. This re-thinking was stimulated by the loss in momentum of the Green Revolution technologies in wheat and rice, and by growing budgetary and foreign debt crises in developing countries which caused governments to look for alternatives to traditional investment projects. There is a growing awareness of the need to improve the effectiveness of support services and to improve the current pattern of production incentives by taking into account the perspective of the rural family.

Despite the essential contribution of the Green Revolution to the improved food situation in a large number of countries, many areas within these countries and many entire countries lacking widespread access to irrigation have failed to benefit from it. Improved procedures are needed for identifying potential new technologies that can fit into rainfed systems. A farming systems perspective could provide the necessary impetus to focus on priority research areas, and to encourage extension services to work in a productive, collegiate manner with research scientists and the farming population. FAO is currently working on specific FSD methodologies to increase the effectiveness of agricultural support services, and for improving project identification and the orientation of policy analysis.

THE TECHNICAL DISCUSSIONS

1. OBJECTIVES

The Technical Discussions held in FAO Headquarters, Rome, on 15-17 October 1991 were organized to explore experiences of development institutions with the farming systems approach and to examine the institutional changes that appear to be necessary for its successful introduction. Most people think of structural changes when the word "institutionalization" is used. Structural change usually implies concomitant changes in resource levels. Both types of change entail clashes with vested interests and new demands upon limited budgets and upon scarce personnel. For this reason institutionalization was defined as "any change necessary to incorporate a farming systems concept or perspective into the work programme of an institution or organization". This implies a concern with organizational structures, as well as with budgetary and personnel resources, plus methodologies and work programmes, and with re-orientation and training. This definition opens the way to the consideration of alternative adjustments - or "soft changes" - as a more immediate means of incorporating a farming systems approach into the operations of development institutions.

Given the above definitions, the objectives of the Technical Discussions can be summarized as follows:

- to identify the major constraints to the adoption of a farming systems approach in development institutions, including those concerned with policy design, research, extension and other support services;

- to specify priority actions necessary to overcome the above constraints;

- to indicate major directions for future programmes to increase the effectiveness of the farming systems approach.

2. PROCESS AND METHODOLOGY

The Technical Discussions were arranged sequentially to allow the papers to be grouped into five main sections:

1. Issues and options in institutionalization of a farming systems approach
2. Adopting farming systems approaches in educational institutions
3. Applying farming systems concepts to technology development
4. Farming systems, Governments and NGOs
5. Country experiences with farming systems approaches

Each presentation focused upon the following main points:

- a brief description of the essentials of the farming systems approach being introduced, i.e. the perspective, the basic methodology and the major actors and institutions concerned;

- a rapid assessment of the progress made in introducing the approach and the main constraints that have affected this introduction;

- steps that have been taken to ameliorate these constraints and any future plans to improve the current situation.

In preparing these proceedings, the papers have been edited and grouped as indicated above, each preceded by a synthesis which attempts to convey the importance of the subject to agricultural development and to highlight the major problems and solutions pertaining to the institutionalization of the aspects of farming systems covered by that area. The institutions and actors that would probably be involved were identified, together with the main methodologies that should be introduced.

It is hoped that the distillation of the experience brought together in Rome will result in programmes of positive action. This should accelerate the application of the farming systems concept to agricultural development activities in the widest possible range of professional functions and development institutions.

II

ISSUES AND OPTIONS IN INSTITUTIONALIZATION OF A FARMING SYSTEMS APPROACH

Overview and Synthesis

A glance through the development literature will reveal that there is a range of development approaches which incorporate the term "farming systems". Some aspects of their core concepts and some components of the methodologies that are employed, have been known and applied for many years. Certain systems approaches have evolved from earlier ones, so there are many similarities between them. A certain amount of confusion concerning the definition of different approaches exists, therefore, together with a lack of clarity about their precise aims and the techniques which they utilize.

However, it is generally agreed that all farming systems approaches have the following common denominators:

- an emphasis on the farm-household that facilitates a bottom-up view of development,

- a holistic and integrative perception of a given situation that stresses interactions and linkages,

- inter-disciplinary work methods that combine the insights and viewpoints of a range of disciplines including those from the social sciences,

- a recognition of the essential role that community participation plays in the development process,

- a problem-solving methodology that attempts to raise the living standards of rural inhabitants rather than simply solving a specific technical problem or achieving an isolated production target,

- the realization that different, often conflicting, sets of objectives must be reconciled.

The most common form of systems approach concerns the **generation of agricultural technology using on-farm testing and with farmer participation.** Practitioners of this Farming Systems Research (FSR) approach have often failed to secure collaboration from extension workers, or from those concerned with other agricultural support services. FSR has often been a top-down process with farmers confined to the role of observers. Farming Systems Development (FSD), promoted by the FAO, seeks to improve this situation through participation and addressing the whole range of agricultural development problems and opportunities during both the analysis and programming phases of any development actions.

Whether the systems approach chosen is sharply focussed as in FSR, or more comprehensive as in FSD, it has rarely been adopted on a wide scale by the agricultural development fraternity. This has given rise to an examination of the factors that are responsible for this failure to institutionalize the systems approach. Major problem areas have been identified. These mainly concern the attitudes of development practitioners, the nature of development institutions and the adequacy of the methodologies applied by the various systems approaches.

With regard to individuals, most practitioners have been trained in the classical scientific method - the so-called reductionist approach - of concentrating upon a single variable while holding all others constant. This is the very antithesis of the holistic approach that takes all relevant interactions into account. Few practitioners have any degree of sensitivity to fundamental rural linkages, inter-relationships or interactions. Their training and experience conditions them to adopt a didactic approach - passing messages to a passive clientele.

The integrated nature of rural life is seldom reflected in the structure of development institutions. These are commonly located within separate ministries and generally focus upon a particular subsector or service. They are mainly staffed by specialists trained to work on a limited aspect of the agricultural process. The situation is exacerbated by the fact that government services tend to be highly centralized.

If the systems approach is to have a wide impact upon the process of agricultural development, it will be necessary to turn concepts into operational procedures. The trap of developing rigid blueprints should be avoided and optimal approaches will obviously vary with the nature of the organization and the task in hand. Nonethless, too much time has been spent on refining concepts and definitions, and not enough on cataloguing successful approaches to specific tasks

In order that the systems approach be widely applied, a three-pronged attack needs to be mounted. Personnel will have to be trained and attitudes modified through re-orientation courses. Incentives to change need to be introduced and specialists who feel professionally threatened need to be reassured.

In the case of of individual institutions, experience with the introduction of other development perspectives indicates that complete restructuring will usually be unecessary. A series of "soft" adjustments, such as networking and modifications to staffing and working methods are available. These modifications will be most successful, however, where radical action is taken by government to decentralize decision-making and budgetary allocation. This process should permeate regular programme structures and not simply be confined to short-term projects. Decentralization also implies greater power being allocated to the community level, with more powers and resources devolving to farmers.

In addition to re-orientation of individual attitudes and adjustments to institutional processes, however, more efforts need to be concentrated on recording and publicizing successful introductions of systems approaches. These efforts should be accompanied by the development of specific work procedures and methodologies to integrate the process of technology generation with its dissemination, adaptation and application. This involves the introduction of a systems perspective into agricultural support services and programming activities, largely by adapting and strengthening existing methodologies.

INCORPORATING THE FARMING SYSTEMS APPROACH INTO THE WORK OF AGRICULTURAL DEVELOPMENT INSTITUTIONS

by

Hall, M.[1] and K.-H. Friedrich[2]

1. INTRODUCTION

In recent years much has been written on farming systems and the concept has been extensively discussed at a wide range of forums. These efforts have yet to be commonly reflected in current approaches to agricultural development. One reaction to this lack of change has been to focus on the perceived problem of **institutionalization** - defined as the adoption of a particular concept by an institution or organization for incorporation into its work and procedures.

Major changes to institutional structures and procedures have often been identified as a precondition for the adoption of a farming systems approach. The paper questions whether this type of radical modification really lies on the "critical path" in terms of the actions needed to ensure that the farming systems concept has a significant impact on development practice.

It also addresses the issue of making the farming systems concept operational through "soft" changes that do not call for institutional restructuring or big increases in resource levels. The paper underlines the need for further development and dissemination of appropriate methodologies, rather than concentrating solely on changing institutional structures and resource levels.

2. SYSTEMS THINKING IN AGRICULTURAL DEVELOPMENT

2.1 The Evolution of the Approach

Interest in the technical and socio-economic aspects of farming systems dates back at least until the beginning of the century, when a considerable literature was developed on the dynamics of peasant farming. It was followed by a wide body of anthropological literature, mainly on African situations, that

---

[1] Research Associate, Harvard Institute for International Development
[2] Senior Officer (Farming Systems), FAO

adopted similar approaches. These early publications underline the importance of household economics, gender roles and the division of labour, and take account of a wide range of interactions within the farm-household system. This approach was later discarded in most agricultural development circles, in favour of narrower, more technical forms of analysis applied to the farm as a business. The earlier, more holistic approach has only recently started to regain popularity, as it is now recognized as a more complete and satisfactory form of analysis for the smallholder situation encountered in developing countries.

The study of biological and physical flows, such as energy and nutrient flows, probably predates the literature on peasant farming. Although it focuses on bio-physical processes and ignores sociological and economic aspects, this area of study is of considerable relevance to agricultural development thinking. Most importantly, it embodies a systems approach that is usually quantified and often expressed in the form of flow diagrams. Its overt recognition of change over time is of particular relevance to an, consideration of the sustainability of farming systems.

For the past 15 -20 years, the most active area of farming systems - both in terms of conceptualization and application - has been that of farming systems research (FSR). This originated as an extension of the whole-farm approach to farm management analysis applied to the identification of farm-level production constraints to indicate priority areas for research. It has since acquired a substantial body of additional aims and techniques, including an important component of on-farm research. Theoretically, FSR consists of 2 major aspects:

- the development and dissemination of improved technologies and practices that are relevant to particular farming systems;

- the implementation of appropriate policy and support systems to create opportunities for the evolution of improved production systems and to provide conditions for the adoption of technologies already available.(Norman and Collinson, 1985)

Experience to date, suggests that only the initial part of the first aspect has been addressed with any consistency. Indeed, some practitioners even seem to have forgotten the importance of the initial analytical stage and concentrate only on the organization of on-farm testing.

A third main branch of farming systems thinking - agro- ecosystems analysis - has become prominent in the past decade. This approach is based upon ecological considerations, but specifically considers the interaction of human activities and the ecological environment. The work of Conway (1986, 1987) and his colleagues is typical of this branch of systems thinking, which has contributed a valuable body of operational concepts and techniques. In particular, the approach stresses the importance of understanding spatial dimensions within a system, and emphasizes the time factor in any analysis of a given situation. Its concise list of criteria that categorize a system (productivity, profitability, sustainability, equitability, and stability) is a useful analytical device and its use of simple charts and matrices for problem identification and planning is well-suited to the development of projects when working in village situations among people with a low level of literacy. In addition, the proponents of this approach were among the pioneers of techniques for rapid rural appraisal (RRA). Nonetheless, agricultural ecosystems has been used mainly as a research tool applied to the analysis and description of particular situations, rather than being applied by to actual development programmes .

FAO's Farming Systems Development (FSD) represents an attempt to incorporate the various strands of systems thinking into an approach that can be applied to most aspects of agricultural sector development. As such, it is intended primarily as an operational approach concerned with improvements to both working procedures and the relevance of outputs. It is intended for application to a wide range of agricultural development activities. These include research and extension, as well as other support services such as marketing and credit. The approach is also meant to address all stages of the project cycle and many facets of policy analysis.

2.2 The Basis of Farming Systems

The farming systems approach is based upon a holistic view of farm-households. These are envisaged as comprising distinct domestic, farm (crops, trees, livestock), and non-farm/off-farm earning components that both complement and compete with one another. Farmers are recognized to be both producers and consumers, who make rational decisions within the limits imposed by conflicting objectives, restricted resource levels and limited technical knowledge. The direct effects of, and interactions between, various aspects of the external environment are also accorded primary importance. These consist of complexes of physical, socio-cultural, institutional and economic factors. Existing practices are interpreted in this light when attempting to understand their rationale. The application of the approach further implies that any official initiatives also take these same factors into account.

The most frequent focus of analysis is either the community or the individual farm-household system. This latter system can be thought of as the "building block" of a series of systems, at different levels of aggregation, that run from the micro to the macro level and can be visualized as a series of concentric rings emanating from a central farm - household. This array of systems includes the local community, the agro-ecological zone, the region and, at the highest degree of aggregation, the nation. The associated analysis should include any interdependencies between the different units in the array, taking account of the fact that their boundaries are not fixed and can be varied at will, according to the type of analysis to be undertaken. In this respect there is a direct parallel in the choice, by FSR/E practitioners, of different "recommendation domains" according to the commodity being focused upon.

A further important systems consideration relates to the "vertical" stages in the process of production and consumption - from the manufacture and acquisition of inputs, through the production and storage stages to those involving marketing and transformation. At each stage of the process that is examined, and at each level of aggregation being considered, the interactions between the system and its external environment are analyzed. This analysis, together with that of other dimensions of the system, is further complicated by considerations of risk and uncertainty at all stages.

The application of a farming systems approach usually involves several overlapping stages, i) initial survey and analysis of the system, ii) discussions and planning, iii) implementation of necessary improvements, investments and research activities; and, iv) monitoring and evaluation. An initial analysis should highlight major problems and constraints, conflicts between special interest groups and between competing objectives within each group. Conversely, it should generate a listing of strengths and opportunities, leading to the formulation of alternative approaches to improving the current situation. Analysis should include an examination of interactions with the external environment, as well as between individual components of this environment. In some cases, such as a review of an extension service or credit agency, it could concentrate upon the relationship between the core system and one particular component of the environment - in this example the institutional component.

Subsequent to an initial analysis of a particular system, the improvement procedures that are adopted will take different forms according to the job at hand - no fixed set of planning and implementation steps can be laid down. In some cases, any change must be preceded by a prolonged research phase, in others development activities could proceed in parallel with further research and development. **The farming systems approach is best thought of as bringing a new perspective to existing tasks or situations,** thus facilitating the identification of improvement measures. A basic procedure for generating this perspective is to view the situation through the eyes of the individual farm family, as well as through those of the community. This procedure involves community participation at each of the stages described above. A parallel procedure is to analyze the situation from several viewpoints and then to synthesize these in a way that leads to effective solutions to the major problems that have been identified. The first stage of this latter procedure calls for a multidisciplinary approach and the second requires the different specialists to work in an inter-disciplinary manner.

3. IS INSTITUTIONALIZATION NECESSARY?

3.1 Generalist Farmers and Specialist Technicians

It is obvious from the brief description of how it evolved and what is involved, that the farming systems approach is essentially holistic and, hence, inter-disciplinary. If agricultural development is viewed within the context of rural development it is also multisectoral. Farmers and their families instinctively operate in a way that is both inter-disciplinary and multisectoral - this is simply a reflection of what they must do to obtain their livelihood. Farmers have family, social and civic responsibilities. They must practice a wide range of skills within the community and the household, as well as on the farm. Additional skills are needed to generate non-farm and off-farm income earnings. As farmers they must be familiar with both crop and animal husbandry and be competent in marketing, in storage of produce and treatment of pests and diseases; they must manage hired labour, decide on land allocation, arrange credit, weigh likely profit against risk, and perform dozens of other categories of task in the course of the farming year.

The integrated nature of rural life is seldom reflected within development institutions. These generally focus on a single sector (eg.agriculture), subsector (eg.livestock), or upon a particular aspect of development (eg.credit). Institutions dealing with agricultural development are commonly dispersed between several different ministries and parastatal agencies. Most are staffed by specialists trained to work on single components or sub-components of the agricultural process, and educated to think in a "reductionist", scientific manner rather than a holistic, informal manner.

Only a few of these specialists will have any degree of sensitivity to fundamental rural linkages, inter-relationships and interactions - especially when they involve socio-cultural factors. Their training and experience conditions them to remain strictly within their discipline and to adopt a didactic approach - passing limited messages, or technical advice "packages", to an essentially passive clientele. They are seldom taught how to listen to farmers' opinions, and rarely attempt to understand their perceptions, or to learn from the store of knowledge accumulated within the community over generations. Most development workers - and certainly most policy analysts - have extremely limited contacts with anyone in the community of smallholders, apart from family connections.

3.2 Institutional Restructuring

The above structural and attitudinal problems represent a challenge to the incorporation of a farming systems perspective into the work of agricultural development institutions - a process that is generally referred to as that of **institutionalization**. A holistic concept, incorporating a high degree of people's participation, is not easily adopted and applied by sub-sectoral ministries and single-function parastatals staffed by professionals with a narrow, didactic approach. Given these problems, it is evident that changes are required in order to incorporate the new approach. In this sense institutionalization is definitely an issue. Moreover, faced with the enormity of the problem, it is tempting to advocate a radical solution involving a major restructuring of the institutions involved.

The changes involved in this form of institutional restructuring would, however, probably be infeasible in the vast majority of situations. Both for practical and political reasons, established institutions are extremely resistant to changes that involve shifts in power and responsibilities.

3.3 Experience with Institutionalizing Development Concepts

Some of the problems likely to be encountered in restructuring can be seen by examining experience with the integration of other development perspectives. The case of Integrated Rural Development (IRD) is particularly relevant in this respect, since it incorporates a holistic, inter-disciplinary approach and a strong component of people's participation. Experience with the institutionalization of IRD has been disappointing. Faced with the impossibility of creating a single ministry to cover all government business related to rural development, many countries have opted to create a Ministry of Rural Development. In most cases, these new ministries have been starved of manpower resources and development budget and have remained peripheral to the main development processes in rural areas. They have sometimes been given the responsibility of co-ordinating IRD projects that are externally funded, but have lacked sufficient authority to integrate the programmes of other ministries. This has sometimes led to the creation of parallel administrative structures in project areas, that have been impossible to replicate in the absence of external funding, or to sustain after this funding ends.

A less radical approach, adopted in the case of both women's issues and environmental considerations, has been to create a special cell or branch. This solution can work when dealing with a single dimension of development, as in these two cases, rather than with the introduction of a comprehensive, new development perspective such as IRD or FSD. Providing that the new entity is properly staffed and has sufficient resources, staff re-orientation and training can be undertaken, as has often occurred in the case of Women In Development. This specific task is, however, not a permanent feature of an organization, so it could probably be more efficiently allocated to consultants or to outside agencies. The new entity could alternatively be assigned a programme and project appraisal role. This is a permanent function of development organizations and provides a logical point in the operational process for the monitoring or the enforcement of a particular development approach. However, monitoring and appraisal procedures encompass a wide range of issues, including adherence to government policies, budgetary control and cost-effectiveness, and are probably more efficient when centralized. The role of the new entity could, therefore, be better performed by simply including staff with the necessary skills within an existing central appraisal mechanism.

A further example of attempts to institutionalize a new approach is provided by the case of Farming Systems Research (FSR), although it could be argued that in its current form FSR more closely resembles a coherent set of techniques than a new development concept. This experience has been well documented and the consensus that emerges is that FSR has been accepted only: a) where it complements, but does not supplant, conventional research procedures, and b) where additional resources are obtained to fund the new programme - frequently by means of external donor funding. It could reasonably stated, therefore, that, in the sense of the majority of scientists thinking in farm-household systems terms, FSR has never really been fully integrated into the operations of research stations. Rather than influencing a broad spectrum of research work, it has only been allowed to exist where it has confined activities to a limited sphere.

The above examples indicate that, while the initiation of separate new entities has undoubtedly had some positive effects, they have not been particularly successful in achieving their fundamental goal of institutionalizing new development perspectives[2]. In many cases, the change has simply led to the creation of an irrelevant, underfunded service at the periphery of the institutional mainstream, or to serious duplication of mainstream tasks, dilution of authority and procedural confusion.

A major exception to this conclusion is afforded by current efforts to decentralize decision-making and encourage the play of market forces in countries that have previously operated centralized, command economies. In these cases, policy re-orientation must be accompanied by fundamental institutional reforms if there is to be any economic improvement.

3.4 Alternatives to Institutional Restructuring

Despite the fact that most discussions on the subject of applying farming systems approaches have revolved around the modalities of restructuring institutions, the experiences discussed above indicate the need to develop a modified approach in many cases. Because of the inherently holistic nature of systems analysis, inter-disciplinarity is a key factor in the farming systems approach. Inter-disciplinary working groups are much easier to organize at the field level than in headquarters. In the field, personnel from different institutions live and work in close proximity and this has a powerful effect in breaking-down institutional barriers. The creation of such groups, even at the local level, is extremely difficult where responsibilities are split between ministries and where

---

[2] In contrast to new types of units for executing specific new functions such as Monitoring and Evaluation, or providing specific new services such as Global Information Systems.

centralized work programming and budgeting is the norm. It is likely, therefore, that a precondition for widespread use of inter-disciplinary approaches is a decision to decentralize the administrative process. Once this fundamental change has taken place, however, inter-disciplinary working methods would call for adjustments to existing institutional arrangements, rather than for their complete transformation.

At the ministerial level, different departments habitually work in isolation and the problem is exacerbated when these separate departments become separate ministries. A strong ministry, such as a Ministry of Finance and Planning, can sometimes provide the secretariat for a co-ordination mechanism that could involve frequent meetings of senior representatives from all relevant ministries. Day to day holistic analysis will, however, almost certainly require the secondment of specialists between ministries (eg. an livestock husbandry specialist to a Crops Ministry). The approach will also necessitate the recruitment of specialists in new disciplines, such as sociology and economics, into traditionally technical and scientific strongholds.

Community participation and farm-household analysis involves a complete change in skills and attitudes on the part of most staff. The requisite sensitization, re-orientation, and education will have to be organized on a vast scale - almost all employees will need to be involved. However, this process can be carried out in stages. It can, for example, focus initially on one of the major functions of a large ministry - perhaps the extension or research services (there is a great advantage in tackling these two in parallel however), or by introducing holistic, systems criteria into the project appraisal mechanism of the institution concerned.

The following changes adhere to the spirit of the preceding discussion by focusing on "soft" adjustments - incremental initiatives that do no involve radical restructuring of institutions to begin the process of institutionalization:

A. Education and Training

Diploma and degree training in FSD
Orientation Workshops
In-service Training Seminars
Training Materials (Manuals, videos, cassettes)

B. Operations and Procedures

Operations and Procedures Manuals
Appropriate Project Appraisal Criteria
Adjustments to Budget Allocation Procedures
Strengthened Monitoring and Evaluation

C. Staff Adjustments

Secondment of specialists within ministries
Secondment of specialists between ministries
Recruitment of new skills (eg.Ecologists, Sociologists)
Hire of specialist consultants
Cooperation with NGO's and farmers' organizations
Cooperation with educational/research institutions

D. Networking

Coordination by a specific cell in the Ministry of Planning or Office of the President
Coordination through regular inter-ministry meetings
Common data bases (projects, planning statistics)
Inter-agency Task Forces
Membership of International Networks

4. FROM CONCEPTS TO TECHNIQUES

4.1 Why Farming Systems Approaches?

The preceding section is concerned with adjustments to working methods, information flows and organization of human resources that would facilitate the introduction of the farming systems approach. Nonetheless, the paper has so far avoided the fundamental question of whether the introduction of a farming systems perspective would assist the process of agricultural development. Is it possible that farming systems is simply a concept without applicable methodologies? Could it be that its holistic message of "everything being related to everything else" is both theoretically self-evident and operationally useless? If this is really the case, there is nothing to institutionalize!

The above queries are not totally unfounded, since the proponents of the farming systems perspective have, with few exceptions, been content to remain at the theoretical level without tackling the issue of operational implementation; thus answering the fundamental question of, "so what"? Any practical means of incorporating farming systems analysis into the operations involved in agricultural development has to start with this question. Only by identifying exactly what are the implications of the concept - if any - for the operations that must be performed, can change be induced. Once these implications are understood, appropriate methodologies can be selected and working procedures can be altered in a pragmatic, incremental manner, within existing institutional structures.

The incorporation of a farming systems approach is usually justified on the basis of an improved understanding of grass roots situations. This understanding is assumed to be gained through the adoption of a holistic perspective that takes account of farm and household interactions within the context of the rural environment. There is little doubt that these assertions are true; it is also true that the dimension of community participation is essential to sustainable development. These statements indicate the potential for certain changes in working procedures, but do not provide enough material to justify radical changes in institutional structures. Clearly a great deal of work remains to be done to turn these generalizations into methodologies useful for clearly identified functions within specific types of organization.

4.2 The Fundamentals of Analysis

The importance of farming systems analysis is that it confronts the agricultural development worker with the fundamental reality that farmers face each day of their lives - the existence of multiple objectives and alternative production and consumption choices that inevitably imply competition for their limited resources and managerial skills. The agricultural production aspect of this situation is basic to orthodox whole-farm analysis. It has been extended in the farming systems approach to incorporate the wider range of family aims and objectives. Moreover, the analysis of scarce resources and other production and consumption constraints has been elaborated to include a systematic consideration of all aspects of the environment (technical, ecological, socio-cultural and institutional). Despite the need to balance a range of competing objectives - many of which are difficult to quantify - the analysis of alternative uses (opportunity costs) for resources, is the key to understanding decisions concerning both supply and consumption; whether they are made by farm families or by governments.

Until the implications of multiple objectives and scarce resources - and consequently of opportunity costs - are fully comprehended by all relevant staff, the farming systems approach has not been completely adopted. As a result, most technicians will continue to give advice, or develop new technology, on the basis of technical feasibility and physical productivity increases or, at best, after applying a simplistic profitability analysis that omits any consideration of alternative uses for scarce resources that can not easily be assigned a price, eg. family labour. In addition, even where profitability is taken into consideration, the simplifying assumption that the sole objective of the farm family is to maximize profit does not reflect the true situation. To complicate the situation still further, the aggregate effect of individual changes is usually ignored. This can lead to serious shortages of inputs or to the collapse of the market for specific outputs. The potential range and sequence of diagnoses and prescriptions can be summarized as follows:

- not technically feasible;
- technically feasible but not financially profitable;
- financially profitable but not economically profitable when the opportunity cost of inputs are considered;
- economically profitable but not accorded a high priority;
- high priority but potential profit outweighed by risk;

- otherwise sound but could transgress socio-cultural norms or exacerbate group tensions;
- otherwise sound but aggregate effect of adoption could lead to shortages of inputs or collapse of product markets.

Most technical specialists seek only partial solutions that seldom progress beyond the first step of the above sequence. In many cases, this leads to sub-optimal changes, or may even cause serious distortions in resource allocation patterns. Only in rare cases is an optimal schedule of changes planned, since the repercussions of a change on the rest of the system, or upon larger aggregates of the system, are generally ignored. The actual situation is further complicated by tensions between individual actors or by different social groupings and by risk and uncertainty arising from specific components of the environment, or from interactions that occur spontaneously between these components.

4.3 What Techniques are Useful?

The ability to see the "big picture" helps specialists to avoid many pitfalls, but this comprehensive appreciation of a given system requires the adoption of appropriate work methods and analytical techniques. The most basic of these is the **systematic inclusion of the farming population at every stage of a development process.** This step alone would eliminate many misconceptions, wrong assumptions and resources wasted on infeasible or low-priority initiatives. The second most important improvement in work methods is probably the organization of inter-disciplinary working groups. This will serve to broaden the spectrum of analysis and to generate a wider range of proposals for improving a given situation.

An active partnership with the farming population and continual discussions between farmers and a range of specialists working in a collegiate manner, will ensure that development workers begin to appreciate the analysis of basic interactions and the implicit weighing of costs and benefits that lie behind most decisions made by farmers. Further insights can be obtained by using simple techniques such as simple charts, maps, diagrams, decision matrices, cropping calendars, cash flow diagrams, and actors' analyses, which form part of the methodology of rapid rural appraisal. Weighting and prioritization of improvement criteria and the sequencing of solutions are other examples of simple and effective techniques that can be easily adopted. These could provide a set of methodologies - a basic "toolkit" - suited to the educational level and abilities of most field staff. The techniques have the added advantage of being readily understood by the farming population and are, therefore, ideally suited to participatory development approaches.

The above techniques, plus others such as partial budgeting, can, however, only yield insights into limited aspects of a system - snapshots of single facets of the big picture. The whole situation can, admittedly, be partially comprehended in a qualitative way by lengthy discussions with the population concerned. It cannot, however, be quantified through this means and farmers are usually unaware of the details of sectoral and macroeconomic issues. Attempts to simulate the dynamics of whole systems could, therefore, form a major supplementary approach to the methodologies used. Recent breakthroughs in the cost and capacity of microcomputers, together with software improvements, make it possible to build data bases that reflect the major characteristics of many aspects of a system's environment. This type of model can be used by suitably trained staff to explore different scenarios concerning risk, conflict and variability over time within a system, thus strengthening the holistic analysis that lies at the core of the systems approach.

It may be objected that the above suggestions introduce a high degree of complexity into the analysis of systems. No apology need be made for this, however, since it is an inescapable fact that all rural systems are highly complex and any attempt to explore them quantitatively must reflect this complexity. Furthermore, the models are intrinsically holistic and capable of being adjusted to take account of better data, new technology and different weightings for competing objectives. They can provide the underlying analysis for a broad spectrum of agricultural development tasks and many simpler, partial techniques can be applied on the basis of the overviews that they provide.

4.4 What Methodology for What Operations?

It is important, to realize that even a radical new development perspective - such as that provided by the farming systems approach - will probably not involve a revolutionary change in basic work functions, nor will it yield a completely new pattern of work outputs. The introduction of farming systems analysis would still leave the main tasks of agricultural development virtually unchanged. Support services will still perform their traditional functions, new technologies must still be developed and adapted, project analysis will continue to be carried out, controlled prices will be adjusted and other policy measures will be attuned to the developing demand and supply picture as before. However, many tasks will be approached differently. When using the approach, the participation of the farming community is a pre-requisite for all development initiatives, new techniques will be applied to capture a holistic perspective, new types of information and data will be needed (especially at the farm and community level) to apply these techniques and more interdisciplinary work groups will be required. **What** is being done will essentially be the same in terms of intended outputs although qualitative changes will lead to new consequences. **Where** it is being done will certainly change in the direction of more on-farm activities. **How** it is being done and **who** does it will be revolutionized by new greater community participation and by the use of inter-disciplinary teams using improved analytical and planning techniques.

The challenge of adopting a farming systems perspective can only be met if people and techniques are incorporated into appropriate working procedures. Despite important shortcomings, that have previously been referred to in this paper, FSR can claim to have developed well-documented working procedures for improving the technology development process. FSD has documented steps for farming systems interventions at the level of the community. Agro-ecosystems proponents have also described intervention procedures at this level, but the approach from either source is cast in terms of a specific project intervention. What is missing are working procedures for integrating a farming systems approach into the regular work programmes of such services as marketing, credit, extension, and mechanization, as well as in units such as land use planning, project analysis and sectoral planning.

These procedures will obviously vary with the task at hand. However, for each main area of agricultural development a series of basic questions can be asked in order to highlight the necessary adjustments. Questions will focus on the following areas:

- what are the main tasks to be performed?
- what analytical and planning techniques should be applied?
- what skills is it necessary to assemble for each task?
- what form of communal participation is required?
- in what sequence should the techniques be applied?
- what outputs are needed?
- what skills acquisition, training and reorientation is necessary?
- can the above actions be synthesized into specific work procedures or methodologies?
- can the methodologies be reflected in job descriptions and operations manuals?
- what problems and constraints are likely to be encountered in introducing the changes (including an analysis of likely allies and opponents)?.

The prescription contained in this section of the paper could more accurately be described as that of operationalization rather than institutionalization. Only by turning concepts into operational procedures will farming systems thinking penetrate into agricultural institutions and in the process become **institutionalized.** This task should receive the highest priority from all those concerned with applying the farming systems approach to agricultural development problems.

REFERENCES

Conway, G.R. (1986), "Agroecosytem Analysis for Research and Development". Bangkok, Winrock International

Conway G.R., McCracken J.A. and Pretty J.N. (1987), "Training Notes for Agroecosystems Analysis and Rapid Rural Appraisal". Sustainable Agriculture Programme, IIED, London.

Dixon J.M. (1990), "Ways Forward for the Farming Systems Approach in Asia". 1990 Farming Systems Research and Extension Symposium, Asian Institute of Technology, Bangkok, Thailand.

FAO (1990), Farming Systems Development - Guidelines for the Conduct of a Training Course in Farming Systems Development. Food and Agricultural Organization of the United Nations, Rome.

Frankenburger T.et al, (1988), "Identification of Results of Farming Systems Research and Extension Activities: A Synthesis". in: University of Arkansas and Winrock International Institute for Agricultural Development (eds.), Proceedings of Farming Systems Research Symposium, 1988. "How Systems Work", Fayetteville, University of Arkansas.

Friedrich K.H. and Hall M. (1990), "Developing Sustainable Farm-Household Systems: The FAO Response to a Challenge. in "Farm Household Analysis, Planning and Development", Proceedings of a Caribbean Workshop, J Seepersal, C Pemberton and G Young (eds.), The University of the West Indies

Merrill-Sands D., Ewell P., Biggs S., and McAllister J. (1989), "Issues in Institutionalizing On-Farm Client-Oriented Research: A Review of Experiences from Nine National Agricultural Research Systems". Quarterly Journal of International Agriculture, 28(3/4): 279-299.

National Environment Secretariat et al, (1990), Participatory Rural Appraisal Handbook: Conducting PRAs in Kenya. , World Resources Institute, Center for International Development and Environment

Norman D. and Collinson M.(1985), "Farming Systems Research in Theory and Practice", in: Agricultural Systems Research for Developing Countries, Proceedings of an International Workshop, ACIAR Proceedings No.11.

Norman D.and Pratt D. (1990), Harnessing Farming Systems Concepts and Methods for a Community and Area Development, Working Paper, Food and Agricultural Organization of the United Nations, Rome.

THE INSTITUTIONALIZATION OF FARMING SYSTEMS DEVELOPMENT

by

A. Shepherd[1]

1. INTRODUCTION: THE ADMINISTRATIVE IMPLICATIONS OF THE FSD APPROACH

The characteristics of the FSD approach of importance to its organisation are as follows. It is:

- diagnostic: it listens to farmers and others involved in the evolution of farming. It treats farmers as important resources and movers in that evolution. FSD is a learning process not just a set of techniques for project identification.

- a "bottom-up" approach, starting at the farm household level, and seeking to influence the policy and institutional environment of the farm household such that farm households can achieve their objectives.

- holistic and integrative: it does not seek to simplify the realities of farm situations to enable a quick fix solution. On the other hand it recognises that FSD for individual farms may be too ambitious and expensive given the limited resources of farm support agencies in peasant economies. So it has developed an analytical approach which enables farm households to be grouped according to shared characteristics. Farm household modelling integrates all relevant aspects, and offers challenges to many separate agencies for solving constraints or meeting opportunities.

- strongly socio-economic in its method, relying on the socio-economist as chief integrative mechanism. Socio-economists are reckoned to have a greater inter-disciplinary capacity, and to be more sympathetic to a systems approach. Ecologists might perform the same function. Generalist agriculturalists are also equipped to lead FSD.

A number of observations can be made about these characteristics. Firstly, the biggest administrative implication is that the FSD approach requires a much

---

[1] Lecturer in Rural Development, Development Administration Group, University of Birmingham, UK

higher quality of work than is presently available in many poor countries' ministries of agriculture and other agricultural agencies including the now omnipresent NGOs. A higher quality of work is sought in: the degree and depth of understanding of farming systems, the quality of analysis of farm improvements, the nature of interaction between the development agency and its farmer clients, the relationship between frontline extensionists and researchers, the relationships between agricultural agencies and other development agencies, and ultimately in the policy process.

Many efforts at institutional reform have been made over the years aiming at similar quality improvements, but these have been heavily constrained by low pay and motivation in the civil service, low levels of educated manpower particularly at the front line, poorly structured administrations with a high degree of centralisation and long chains of command, and the establishment of project units or authorities partly funded and staffed by aid donors to get around these constraints which have been of short lived benefit and have proved unsustainable in the longer term. What is clear from these experiences is that generalised solutions are unlikely and that the path will be difficult.

Secondly, some decentralisation of decision making is a requirement for the implementation of an FSD approach, simply because farmers' objectives are not always the same as the state's or the international agency's. The FSD process can be seen as a method for clarifying farmers' objectives and harmonising them to the extent possible with wider societal or governmental objectives. Because it starts at the farm household level the FSD approach necessarily gives special weight to farmers' objectives. If the requirements for fulfilling these objectives are constantly frustrated by clashes with superior interests, the FSD approach will be a waste of time. There must therefore be some room for manoeuvre, some scope of decision making as close to the farmer as possible. There are a variety of methods to achieve this, reviewed in Sections 2 and 3 below.

Thirdly, while the FSD approach is highly participatory with respect to information gathering and testing of new technologies, the process by which these adaptations or new technologies are generated is not susceptible to control or even influence from farmers. For example, the FSD approach retains a strong emphasis on formal financial analysis requiring quite a high degree of specialist knowledge. "The responsibility for defining the best available technology for improving a farming system remains with the FSD team." (FAO, 1990: 205) Ultimately the farmers' role is to test and approve, and then to adopt, but not to define or to determine. This suggests that a traditional hierarchical research and extension arrangement can adapt to the FSD approach. The main argument of this paper is that if it can adapt, the adaptation required is very substantial and has many aspects.

Lastly, the FSD approach advocated by FAO is catalytic rather than transformative. In practical terms an FSD team is formed in order to advance understanding of the farming system, to identify constraints and opportunities within it and in the wider environment, and to analyze and advocate improvements in technology, organisation, practices or policies. There is a presumption that existing organisations will deliver, if necessary modifying their procedures, structures and behaviour to respond to the FSD team's interventions. In an atmosphere where budgetary, physical and manpower resources are scarce and professional attitudes hardened, this will not be a conflict free process. FAO argues that the FSD team must not try to take over the functions of existing organisations. Ministries of Health have similarly sought over the past fifteen years to advance the primary health care (PHC) approach through catalytic District and other level PHC teams: this not very encouraging experience is reviewed in Section 2.3. This paper argues that in the current era of grand state restructuring plans the FSD approach can afford to be bolder. If FSD teams are to achieve a level of change to balance their costs and to meet the challenges for which the FSD approach has been designed decision making powers must be delegated to them which provides them with resources which they can use to influence other agencies which assist the process of farming system development. A model which would permit this is discussed in Section 3.3.

2. THE INSTITUTIONAL CHALLENGE: INTEGRATION AND MULTI-SECTORAL ADMINISTRATION

There is a variety of experience in development which can shed light on the difficulties and possibilities of developing an inter-disciplinary, multi-sectoral approach to holistically defined problem situations. This section reviews a selection of these. Firstly, the lessons from the long experience of integrated rural development programmes are recapitulated. Secondly, some lessons are distilled from the more recent focus on the role of women in development, where women's interests and objectives have acted as an integrative principle, through an analysis of the Kordofan Integrated Women's Development Programme. Thirdly, the experience of multidisciplinary teams in different development situations is documented to illustrate the difficulties an FSD approach must confront. Finally, the question of whether the project is the inevitable form for the FSD approach is addressed, and the implications of encapsulating the approach in this form are elaborated.

2.1 The Lessons of Integrated Rural Development (IRD)

FSD is not as ambitious in its scope as IRD. It does nevertheless require a significant level of integration in its activities if it is to bear fruit. Professionals needed in FSD will certainly be found in more than one department of the Ministry of Agriculture, may come from two or three Ministries, and may even be employees of local as well as central government.

In such rural development programmes it is rare to find resources needed to carry out the programme under the control of a single decision-making unit. Activities entirely controlled by any responsible unit are few; co-ordination with others is required to release necessary resources of manpower, funds, transport, supplies at the right time and in the right quantity and quality. There will be yet further resources entirely beyond the control or influence of the unit: here it can only appreciate the decisions made by others (Honadle and Cooper, 1989).

A number of studies of IRD project administration were made for USAID around 1980, in particular by Cohen (1979), Honadle et al (1980) and Morss and Gow (1980). This followed a decade of USAID investments. What follows is an attempt to distil lessons from those studies.

These studies highlight the fact that IRD, with its central objective of assisting the rural poor to increase their capacity for positive engagement in the development process, was politically charged and therefore sensitive by nature. It is not yet clear how politically charged FSD is, but if its objective is seriously to assist small farm households to improve their incomes and production it is not uncontroversial in many circumstances, since government rhetoric in favour of small farmers is often not born out in practice. Furthermore, if FSD's approach is to give the farmer the status of expert, this then threatens existing experts. Finally, the conclusion of FSD processes is unlikely to fit easily with conventional views on how to develop peasant agriculture from reorganising land tenure to throwing fertiliser at the farm. Vested interests, professional as well as political and commercial should not find it straightforward to use FSD in their own interests. In all these ways FSD lays down challenges.

For Honadle et al (1980: 202-5) any controversial programme which seeks a measure of integration to achieve its objectives will be very sensitive to how it is organised and managed. The first lesson is that organisation design should be a salient part of any programme evolution. For example, if participation of beneficiaries is required, as it is in FSD, organisational mechanisms for it should be explored and finally specified: participation cannot be left to chance. If it is, managers end up dealing only with pressing day to day problems such as vehicle maintenance, and lose touch with their clientele.

Participative organisational mechanisms are not a straightforward matter, however. "Rural societies are usually stratified and it is often difficult to establish relations with organisations which represent the interests of the rural poor....IRDPs may identify those organisations which are most representative and work through them, strengthening them in the process, or they may try to improve the position of the rural poor within organisations where they are not adequately represented. Another alternative is the delivery of goods and services directly to the rural poor themselves. By-passing local organisations altogether, however, is unlikely to lead to sustained capacity building....Where there are many different people's organisations in an area IRDPs may have to choose between working through one main organisation, thereby facilitating co-ordination and concentrating effort, and distributing their attention among a number of individuals or interest groups likely to benefit in some way from the project...Since the poor often lack confidence in their own capabilities and knowledge of the alternatives available to them, effective participation may require "conscientisation", or awareness-raising efforts." (Conyers et al, 1987: 18)

Organisation design, like the design of IRD in general, should be a process: no blueprint should be set out from the start. Of course a starting point should be identified - for example, a management team, or a project unit - but this should be expected to evolve as the programme evolves. Most writing on IRD from the late 1970s has stressed the need to avoid rigid blueprints, to treat IRD as a learning process; many governments and donor agencies have funded such programmes in the 1980s. A case study of one such programme is presented in Section 2.2 below. The approach has not yet been comprehensively evaluated, however.

What is clear from these efforts, however, is that capacity building has become a key mechanism and focus for IRD (Conyers et al, 1987). Whereas the earlier generation of ambitious IRD efforts tended to set up new organisations to implement a blueprint plan, deliberately bypassing existing administrative and political arrangements which were perceived as inadequate or hostile, the later IRD efforts have been characterised by attempts to work with and within decentralised administrative and political arrangements, improving their capacity for continuous planning through training, strong and appropriate monitoring and evaluation procedures, and a more "hands off" approach to budgetary control. A number of governments and donors have been willing to adjust their administrative requirements to this approach. Some (eg GTZ - German Technical Co-operation) have moved towards long term (15 year) funding, with project frameworks rather than detailed plans as guiding documents. A few donors (principally NORAD and SIDA) are now moving even further towards programme funding, leaving the entire project process in the hands of government. Capacity building then occupies centre stage.

A project which has pioneered the capacity building approach is the Northern Zambia IRDP. This entailed the development of management systems in the District Councils in a learning process to cope with District infrastructure development needs. The project played the role of facilitator. As a result there was no blueprint, no precise costing, no control over planning priorities or implementation; it was not susceptible to convention rates of return analysis; funds had to be provided "up-front" to give districts full responsibility; quantitative evaluation was difficult; the project team had to work very closely with the donor to build its confidence in the approach. The results after 5 years were:

- accurate project costing
- importance of information collection and presentation in defining priorities
- annual three year rolling plans
- departments implementing 210 discrete projects

- setting up of new operational programmes eg ox-ploughing
- development of monitoring systems in agriculture and health
- the involvement of a wide range of district and sub-district representatives and staff in decision making. (Goldman et al, 1989)

One aspect of capacity building which was neglected in the earlier round of IRD programmes was the training of managers to run complex projects. Managers need the skills of organisational analysis, negotiation, networking, team building. They need to be able to adapt the project's organisational design as the project evolves.

A further neglected aspect was that of incentives. In any integrated work the incentives to behave in a dis-integrated fashion are usually stronger than those in favour of integration. An activity requiring integration obliges its managers to review incentives and sanctions to make sure that they encourage it as far as possible. Headquarters support for horizontal integration can be critical: funds and other resources can be allocated or withdrawn on the basis of degree of integration achieved. Promotion prospects could be partly determined by an individual's success in team-working. However, despite widespread recognition of the problem, few organisational incentive strategies have been evolved for IRD. Reliance tends to fall back on personal financial incentives.

While most attention is generally given to horizontal integration, vertical integration can be as or more problematic. It is not unusual to find different levels of a ministry competing to control a programme which has a significant flow of resources. While headquarters support is important, strong headquarters direction and control over resources can vitiate the decentralisation necessary to achieve integration.

A final, and perhaps strongest conclusion from IRD efforts, which bears frequent repetition, is Chambers' recommendation (1974:24 - 26) that co-ordination requirements should be minimised. Co-ordination is costly, in terms of time and recurrent budgets. As far as possible decision processes should work in parallel with as few decisions as possible depending on each other. Information about decisions, on the other hand, needs to be widely shared.

The classic progression experienced by many IRD projects is from initially offering a comprehensive package to later offering a more feasible minimum package; likewise from initially attempting to work with many organisations to later working closely with a few. Partly this is a process of working out what is important in any particular situation and given broad objectives; partly it is a matter of what is feasible and not too demanding in terms of negotiation and inter-organisational relationships.

A far better progression, from single entry points to sets of mutually supportive programme components, relying on co-ordination or integrated activities only where logically necessary, is the approach most commonly adopted by NGOs (eg Field, 1982). It is a low key approach which does not go beyond implementation capacity at any point in time, which encourages innovative organisational approaches as problems are perceived and confronted with scarce resources. This could well suit FSD under many circumstances - where FSD skills are scarce, where it is seen as a controversial or different approach, and where resources allocated to it are few.

2.2 Focusing Development on Women as a Driving Force for Integration: the Lessons of the Kordofan Integrated Women's Development Programme

Focusing a development programme on improving the social or economic situation of a particular social group should have a structuring and integrative effect on the programme. As with farmers, women are likely to occupy one or a set of situations in any particular place, giving them more or less coherent sets of needs and facing them with sets of constraints and opportunities.

Women's development, like FSD, is a threatening activity, seen from the position of male dominated professions and departments. In Kordofan (Sudan)[2] it was found that a small scale, low profile, rather unsystematised approach was best to start with. This avoided any suggestion that the programme sought to take over any existing activities, and avoided suggesting that its approach was more correct than others'. It derived its activities from two sources: (1) the existing proposals or activities of various regional government departments, adapted to suit the specific client group (women), and monitored to make sure that it worked, and that it worked to the benefit of the client group; (2) the logic of women's situation which was revealed through studies (eg of declining fuelwood availability which led to investment in women's woodlots, access to grain which led to investment in community grain stores under women's control etc), through the discovery of resources with development potential under women's command (eg the rainy season milk surplus among cattle nomads, which led to a cheese making extension programme for nomad women, the small farms usually near the house under women's control, which led to an extension programme focusing specifically on that previously neglected farm sub-system), and through intensive and continuous interaction with village women by all agencies involved in the programme. It is possible that this bottom-up, step by step, low profile approach of "anti-method" is appropriate for FSD work in situations where strong vested interests are ranged against the approach.

---

[2] For a complete description and analysis of the Kordofan Integrated Women's Development Programme see Shepherd (forthcoming 1991)

In this Unicef-funded programme the regional Unicef office played the catalytic role. The programme was slow to start, low profile, but supported with adequate funds. Although highly interactive in its approach its objective was to identify and fund fairly simple activities which could be replicated on a fairly wide scale, taking advantage of the tendency and ability of government agencies to spread their work widely but thinly. Bureaucratic administration is not generally able to be very sensitive to inter- or intra-community variation: certainly in Kordofan a more locally tailored approach would have only been appropriate for NGOs to undertake. In Section 3 below it is suggested that a thoroughgoing implementation of the FSD approach beyond the catalytic level cannot be expected of an untransformed bureaucratic administration, and the alternatives are examined.

The catalytic role was possible where regional departments were sufficiently flexible to enter into a continuous process of negotiations on programme development. In one case, a department run by and for women, rigid central control was expressed in adherence to two or three basic ideas about women's development which were paternalistic (!) and out of keeping with the programme's overall approach. Sadly, despite the predominance of women in the organisation, it did not prove possible to develop sensible activities. Rigid central control in this department was perhaps tolerated at regional level because regional managers' education levels were significantly lower than those of central managers. This was not so in other departments. Indeed the degree of flexibility and analytical ability among middle level departmental managers was a key to the overall success of the programme: this must be attributed substantially to their high level of education - a tribute to University of Khartoum degrees.

An expression of the catalytic role was to use evaluation as a management tool. Evaluations were carried out frequently and on diverse issues, both specific and general. This was an extremely useful process of exposing departments whose activities were sometimes cast in stone to interactions with perceptive outsiders. Efforts were also made to encourage departments to evaluate each others' activities. Three key institutional issues were raised by evaluations. Firstly, the question of integration or co-ordination of activities: one proposal was to establish women in development (WID) units in Unicef and in the Regional Planning Department to lead and co-ordinate the programme. A second, more practical proposal was that activities where there were clear logical relationships should be grouped and work closely together at a conceptual and if desirable a practical level. Practically, co-ordination is costly: it should therefore be minimised. This also removes the potentially threatening character of co-ordination, which can easily slip into attempts to control. On the other hand the logic of women's situation in the region increasingly suggested that their participation in economic activities could become a burden to them, and that other burdens (fetching water and firewood, caring for the sick, child care) needed to be removed or reduced if the new economic activities were to be fully beneficial. Thus co-ordination of economic activities with improvements in social infrastructure (primary health care, village water supplies and sanitation in practice) became an important aspect of the programme, made easier by the prominent role in the region of Unicef in funding such activities.

Secondly, the question of capacity building: should the focus be on village or departmental management capacity ? Clearly, for an integrative programme the stronger village management the better since that is where issues are treated already in a holistic fashion. However, in a situation such as that in Kordofan, the capacity of departments to respond creatively to village initiatives, even to facilitate those initiatives in the first place is critical. Over the programme so far one can detect a desirable shift, however, towards village women taking fuller control over programme activities. Departments have been encouraged to let women's groups or co-operatives run their own businesses to a greater extent, to market their own products through channels over which they exert some influence, and to pay full costs for the services they receive. This has happened not for ideological reasons, but for practical reasons: as the production of cheese by nomad women grew, the ministry could no longer absorb it all; the interventions of the Co-operative department in the management of grain store co-operatives was heavy handed and created confusion about the ownership of the enterprise and what could be done with the commodities passing through it; rampant inflation rapidly eroded the value of nominal repayments made after the event by beneficiaries, while at the same time in many cases participants were willing and able to pay full or near full costs of the services rendered. This willingness was consistently underestimated by regional officials.

To summarise the lessons of this programme: (1) A catalytic role in a controversial and experimental programme is best played initially in a low key way, relying on co-operation, and a modest injection of funds, together with a fairly liberal approach to concept development, not too closely tied to any dogmatic method or procedure. It is particularly important to find value in what agencies are already doing or propose to do, and to work alongside them in concept and programme development, using monitoring and evaluation as management tools. (2) This approach would build on the strengths of bureaucratic administration: its ability to replicate simple interventions widely, but would find difficulty succeeding where departments are strongly controlled from the centre. (3) Integration would be sought where logically necessary, and avoided if too painful or threatening. Localising management of the programme as near as possible to the village level is likely to foster integration.

2.3 Teamwork for FSD: Lessons from Multidisciplinary Team Situations

There are a number of common situations where teamwork is critical to programme success. IRD is of course one, primary health care another, and the broad area of district development a third. These will be considered briefly below. Very close to FSD is the multidisciplinary teamwork involved in FSR. This will be left for others' contributions.

2.3.1 Teamwork in IRD

It is frequently observed that teamwork on the ground, close to actual development activities is easier than at headquarters levels. This is perhaps because positive personal relations can play a strong role (though poor personal relations may equally be a barrier), but perhaps also because the imperatives for co-operation are greater: clients may demand it, targets may not be achieved without it, give and take is more likely. Tasks should also be more easily defined, clearly allocated and with less overlap at local level, so long as job descriptions are clear and conceived in relation to each other.

On the other hand conflicts can also easily emerge at local level. These may reflect conflicts at higher levels, but there are also those which derive from misunderstanding or the operation of local vested interests, which have some hope of local resolution. A conflict resolution mechanism - a supervisor, an arbiter, a committee process, a local democratic body - should be in place to deal with such issues.

Morss and Gow's study of USAID-funded IRD projects concluded that there was often conflict between team members, and that in such big projects only a specialised management team or academic contractors had provided the necessary strong leadership (1980).

Since FSD is likely to be based in one ministry, or at the most two closely related ministries (eg Agriculture and Animal Husbandry) the same degree of strong leadership may not be required. On the other hand, there can be strong traditions of conflict and competition between strongly related departments of ministries which run very deep into a country's history. These require a continuous process of communication, and the creation of shared understanding. Under some circumstances this can be more easily achieved by outsiders not raised in the culture of conflict. As in FSR it may be that social scientists/economists are in a better position to lead FSD teams, since they do not have vested interests in any particular technical form of development.

2.3.2 Teamwork in Primary Health Care

Whereas studies of IRD have identified conflict among team members as the major problem of IRD teams, the major problem for PHC teams is the lack of delegation of resources to support a decentralised management process. Ministries of Health typically reluctantly allocate untied money to district health teams or local governments advised by their health professionals. Major budgetary activities - like immunisation, drug supply and special disease control programmes - are kept under close central control, leaving relatively less important activities such as construction of PHC facilities etc to district teams.

2.3.3 District Teams

District Teams are usually intended to co-ordinate development in a district. The incentive to co-ordinate activities of different departments is often weakly established in the absence of a project like the Northern Zambian IRDP mentioned above. Their operations often seem to participants to be empty rituals because they have no financial resources at their disposal, they become arenas for power struggles between departments or between the district administrator and technical departments, accountability may be confused, and central government indifferent to the team's views. The lessons would be that co-ordination should be optimised (or even minimised) rather than maximised; and that resources are required to back up and make meaningful joint intellectual efforts in data collection, analysis, priority setting and so on.

2.4 FSD: Projects versus the Pervasive Approach

FSD is a concept developed by an international agency. To be implemented it will, if it is like other international agency concepts, first have to be cast into the project form for which donor funding can be raised. In its concept it is already amenable to projectisation since it relies on an FSD team to bring about the desired change in approach. The work of an FSD team can easily be expressed in project terms. Even where FSD is taken up enthusiastically by government it is likely to be expressed in project form, since it is innovative and involves specialists from more than one department or professional background. Some governments (eg Indonesia, India) have adopted the project form extensively within their own administrative procedures. It allows the financier (in Indonesia the Ministry of Finance) a measure of control where projects are funded in local or provincial governments which have some autonomy (eg the Inpres Dati II programme). Central control tends to be expressed in terms of standardisation of practice and quality control and is a mechanism for transferring ideas from one place to another rapidly. In donor-funded projects the same scope of influence is possible, although donors' influence is generally severely curtailed by government rules and regulations.

In societies where the civil service is very run down an additional argument for projectisation will undoubtedly be that only through a project will it be possible to pay the incentives which will motivate officials to do the hard and high quality work required in an FSD approach. FSD requires experienced staff to work (and probably live) at local level, enduring the deprivations which that may entail in a poor society. Monetary incentives are probably also required in this respect.

As argued in Section 1, FSD requires a decentralisation of decision-making. In a centralised ministry, like Agriculture in most circumstances, the project represents an opportunity to decentralise some decisions and control over some resources wholly or partly to local level, while maintaining accountability to the centre or a donor.

Genuine decentralisation of power and resources to local government would probably obviate the need to projectise FSD: it would then be a matter of whether a local government was enthusiastic enough about the concept to restructure its operations around it. Agricultural administration is rarely under the aegis of local government, however. Exceptions are in Nigeria, where agriculture is partly a state responsibility and partly local government's, a few states in India (eg Karnataka.

In northern Nigeria, where many heads of local government agriculture departments have recently been trained in rapid rural appraisal methods, and where local governments have a remarkable level of resources to dispose, would be an excellent venue to try a decentralised approach to FSD promotion. This could be linked with the Samaru Institute of Agricultural Research at Ahmadu Bello University, which has had a long history of involvement in farming systems work, and with the Institute of Administration (Department of Local Government) which has been carrying out training in RRA with the assistance of the University of Birmingham's Development Administration Group.

Out of 75 heads of local government departments trained in RRA methods one third have trained their extension staff using local government funds. However, RRA methods have hardly been used due to a number of constraints: lack of transport, lack of personal incentives to work hard, out in the villages, and the difficulty of challenging the established information system (from village headman to emir, and on to local government). The quality of RRA training is also questionable in the local governments - heads of department were not trained as trainers, nor have they been provided with the necessary simplified training manuals to enable extension workers to understand the methods.

The question raised clearly by this innovative project is whether training by itself, without institutional change, can achieve much. Success currently depends on the personality and self-perception of the heads of department. This is clearly not enough to sustain a new approach to local government programming.

Ultimately the objective would be to spread the FSD approach widely within the agricultural administration. Indeed if it is to work initially at local levels, whether through a project or a fully decentralised approach, it will require a measure of support from key departments and offices within the agricultural administration. The concept must be known and understood at least selectively.

3. FARMER-LED FSD: ORGANISATIONAL OPTIONS UNDER STRUCTURAL ADJUSTMENT

Farmers are the users and guardians of their farms. They usually understand their farm systems better than anyone. Their analysis of potential improvements is cautious and comprehensive, and their adoption rapid once an innovation proves its worth. Hence the logic of on-farm testing, an important part of FSD.

However, the FSD approach could with advantage go further than it does in involving farmers, where this is possible. In fact, where farmers have a good understanding of their constraints and opportunities and can articulate them, there is a prima facie case for putting them in command of the process. This is even more strongly the case where FSD is aiming at maximum use of on-farm or locally available rather than external resources. Once again, a much less methodical and therefore externally directed approach to development could be adopted, relying on the knowledge and control over resources of the farmer.

In an era of structural adjustment where Ministries of Agriculture are being asked to review their structures, functions and manpower, with a view to reducing public expenditure as a share of GDP, farmer-led FSD offers an opportunity to rethink agricultural administration from a fundamentalist perspective.

FSD has come about because top-down, technically oriented agricultural development has failed to address the production problems of many areas, especially in small farm economies and complex ecologies. The structures of ministries of agriculture reflect a top-down view of function: indeed they can only be understood in the context of transmitting messages from researchers to farmers, either in the form of advice or regulations. Manpower is located accordingly: well trained at the point of message formulation, and poorly trained close to the farmer, where it is just a matter of transmitting that message. Training and Visit schemes merely sharpen this process.

Ministries of Agriculture typically absorb a large proportion of recurrent government expenditure (10 - 30 %) much of which goes on staff. Substantial proportions of skilled staff are located in headquarters, carrying out paper work, and rarely interact with farmers. A casual study of communications within the Union Ministry of Agriculture in Delhi estimated that 80% of communications were internal to the headquarters office. Headquarters costs are often massive. The payroll is also a very long one, with thousands of low paid, poorly trained extensionists in the field. These are found time and again not to have relevant messages to communicate to farmers - not for any fault of their own, but because of failings of the system to produce useful information, or to motivate them to interact with farmers. T & V can improve the situation, but is unaffordable for most countries in the absence of aid.

If farmers are indeed experts in understanding their own situations, and in diagnosing constraints and opportunities it would make sense to give them the power and resources to develop their own solutions to constraints and take advantage of opportunities.

In order to do this they would have to be able to interact with well trained agriculturalists who have an understanding of farming system development issues. In a farmer-led process this could be achieved in a number of ways. The remainder of this section outlines the options.

In general this paper argues that a farmer-led FSD approach requires that fieldwork is done by the well trained agriculturalists (or FSD professionals) in an organisation, perhaps assisted by local level or paraprofessional extensionists. Graduates and diplomates are thus the key cadre. Getting them out of the offices and into the field in an active advisory rather than a supervisory capacity is critical. Providing them with conditions of work to enable them to stay as near as possible to their workplace - on the farms - is also vital.

A modern, post-structural adjustment, FSD oriented agricultural administration would therefore have a totally different structure to the classic or even T & V reformed organisation. It would have a shallow hierarchy, a team rather than departmental approach to the distribution of specialist skills, very few purely supervisory posts, and a far smaller headquarters staff list.

The hub of the FSD system would be the local (sub-district) level agriculturalist/advisor. S/he would be qualified as a generalist but ideally also have some specialism through further training. S/he would preferably be a degree graduate or at the least a diplomate. Ideally S/he would live at the local level. Any assistants or village level extensionists would work with and under this advisor. This sub-district team would work with farmer groups using the FSD approach.

Each advisor would be linked in a network to other similar advisors in other nearby local areas. They would have regular meetings and varied forms of exchange. Any specialisms among the advisors would be shared in the network. There might be 5 or 6 advisors in a network.

Each agriculturalist/advisor would report to and request assistance from a district agricultural officer (DAO), who would, for example, liaise with researchers in the district and elsewhere about particular problems and requirements, and arrange for researchers to give advice on the management of on-farm trials. The DAO would be a trainer, responsible for continuously upgrading the FSD skills of the advisors and their assistants. The bureaucratic aspect of a DAO's job would tend to become less important.

Each advisor would also be accountable to a local level democratically elected body, or its delegates (eg a board/committee). Ideally this organisation would be the budget holder, and would allocate its resources according to its own development strategy both between agriculture and other sectors as well as between activities. Initially it might be given only partial budget authority, and elected representatives might sit together with appointed representatives until the organisation's decision making capabilities improved. This type of organisational arrangement is considered further in Section 3.3.2 below.

FIGURE 1a: AN AGRICULTURAL AGENCY'S STRUCTURE TODAY

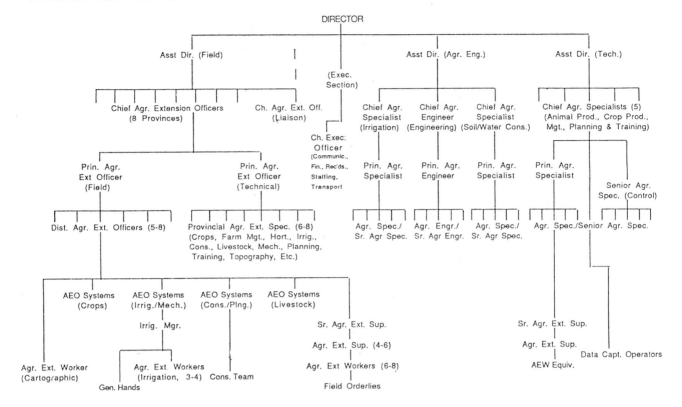

FIGURE 1b: AN AGRICULTURAL AGENCY'S STRUCTURE REFORMED FOR FSD

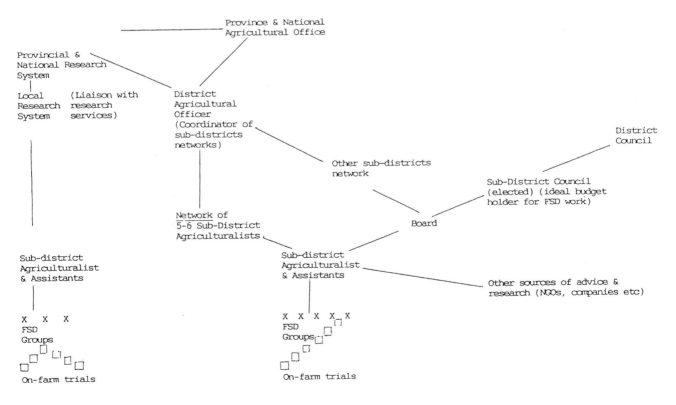

Budgetary autonomy implies that the sub-district body could decide to spend its agricultural budget seeking advice or other services from sources other than the Ministry of Agriculture, such as NGOs or the private sector. In the phase of team building for FSD it might be desirable to prevent this, by tying a part or all of the budget to the purchase of services from the FSD team. Ultimately, a healthy situation would require that farmers be free to purchase services from wherever could best provide them. For most poor countries and most small farmers this would remain the public sector for the foreseeable future.

Beyond this district level system the FSD approach would require a support unit. This could be an agricultural college or university department, a special FSD unit, or possibly an FSR team. Any support unit should be skilled in FSD diagnostic work and on farm trials, or in the case of a college, be rapidly developing the skills.

3.1 Farmer Managed Research

There has recently been an upsurge of interest in involving farmers in agricultural research to the extent of letting them manage on-farm trials and research programmes and/or letting them decide on what topics should be researched and how. This was extensively reported in Chambers and Pacey eds (1989). Their conclusion was that on-farm trials represented only one possible entry into the systems approach, and that good on-farm work is very dynamic. A learning process is involved which may take researchers far away from conventional research methods. Involving the farmers in constructing the agenda for research is especially important. Participation did not imply "asking farmers to approve our experiments, but eliciting their experiments and designs." (Ibid: 160-1) Farmers want immediate results, results which take into account the important interactions on the farm, and they are more interested in participating in research which lies within their own capacity rather than conventional high-input solutions.

Chambers identifies three conditions for the success of "farmer first" strategies: decisions need to be decentralised, together with the resources needed for interaction with farmers - especially mobility; staff need to learn respect for farmers' innovations and develop information systems to respond to farmers' demands; and staff need to be rewarded for work in the field, pioneering new methods, and networking among themselves (Ibid: 186-9).

In Europe Norway's small farm economy provides an interesting case of the rapid development of farmer managed research since 1975. There are now about 100 "research circles", 21,000 members and 400 employees. The system provides researchers who interact with groups of farmers in series of on-farm experiments investigating issues which either the farmers or the researchers think are important. Employees are expected to advise as much as to research. Currently farmers are very keen to cut costs of production as the government is determinedly cutting farm subsidies and thus gross farm incomes. In the Nordosterdalen area the use of slurry to substitute for chemical fertiliser on grass was the farmers' research priority.

Farmers subscribe to join a group, and this funds a growing proportion (now about 30%) of research costs as government reduces its contributions. The subscription represents a small proportion of a Norwegian farmer's income, which is of course substantial by comparison with small farmers in developing countries. As this form of research expands the Ministry of Agriculture proper has less and less to do with extension work, and is increasingly a regulatory organisation, particularly concerned with environmental issues. It is doing less and less applied research, leaving both research and extension increasingly to the farmer managed research circles.

The groups meet frequently in the winter when the nights are long and there is less farm work, and the researchers get on with field work during the growing seasons on farmers' fields but organising the work themselves.

A non-governmental parallel in the gardening field is the UK's Henry Doubleday Research Association (HDRA), which provides a research and advisory service on organic gardening to its 16,000 members. Members have an input through a democratic structure in deciding what the organisation as a whole should research, and then participate in members' experiments, communicating their results by post. The HDRA is about to expand its advisory service to developing countries: it could more usefully extend its organisational approach to developing country NGOs.

To what extent can farmer managed research become a reality in poor countries among small farmers ?...

What are the administrative implications of farmer managed research ?

3.2 Farmer led NGOs

There are two types of non-governmental organisation which may be useful in promoting farmer-led FSD. The first consists of "membership" NGOs which service a defined membership and which are managed by that membership, directly or indirectly, more or less democratically. Examples would include co-operatives, farmer managed irrigation systems. Village councils and committees could also be mentioned here, though these shade into the governmental arena which will be discussed in 3.3 below.

The second consists of "service" NGOs which provide services to the public or a section of the public but which are funded and controlled externally.[3] While many service NGOs provide agricultural extension and inputs, this is usually done in a conventional top-down manner. The NGOs working in the broad field of "sustainable agriculture" on the other hand, have begun to move towards an approach which resembles FSD.

3.2.1 Membership NGOs

In many ways membership NGOs - run by and for members - are the best to co-operate with, being closest to the farmer. However, the scale of some of these organisations (eg co-operative societies and movements) may render them bureaucratic and distant from their members. Where government control of co-operatives is strong their representative function will be limited.

It may be more relevant to work through very local organisations. However, these are often fragile. Furthermore their often egalitarian principles may work against the idea central to FSD that farmers should be categorised according to their characteristics, and recommendation domains developed.

3.2.2 Service NGOs

Compared to government organisations service NGOs are supposed to be able to reach the poor, to be flexible in their approach so that it is adjusted to local conditions, and to be able to promote local autonomous development organisations. This is their much vaunted "comparative advantage".

The work of these service NGOs can suffer from a lack of system, however. Writing about the Gwembe Valley Agricultural Mission (GVAM) in the Zambezi Valley of Zambia, Copestake concludes:

"Stengths" include:
(a) first-hand knowledge, of local production constraints as well as the opportunities presented by the lake, arising from the fact that its staff live and farm in the area,

(b) frequent contact and good working relations with farmers over a wide area through the ten pre-coops and the wide range of activities it carries out with them.

These strengths have enabled GVAM to absorb new ideas, test them and

---

[3] For an elaboration of this distinction see Fowler (1990)

either reject or disseminate them along the lakeshore very quickly. However, GVAM also has weaknesses, some of them born of this very flexibility, including:

(a) a lack of formal planning, recording and analysis of experiments and experience.

(b) a tendency for agricultural extension and consultation to be put off by more urgent demands on staff time, arising from other aspects of GVAM's programme" (Copestake, 1990: 29 - 30).

Clearly, an FSD approach would fit in very easily here, and would help the organisation become a bit better organised.

In Kenya, the Kenyan Institute of Organic Farming (KIOF) is promoting low external input methods of farming in the central highlands. There has been a tremendous rapid growth in demand for its services, especially from small farmers, men and women, who cannot afford cash inputs. It works with self-selected groups of farmers. Up to now it is purely an advisory service, but encourages members to experiment with innovations on a small scale before they take them up widely on their farms, and also to share experience across the group. As it matures, and gains knowledge of what works under different circumstances, the FSD approach would help it to analyze and systematise that knowledge.

The likelihood is that the FSD approach would complement what KIOF is already doing, and would not require great changes in its structure or method of working. It would merely require some systematisation in the process of experimenting and on-farm testing.

There are many other examples of service NGOs already providing services within the sustainable agriculture framework which could benefit from using the FSD approach.

In the UK, the Elm Farm Research Centre combines a research focus and advisory work in organic farming. As with most modern advisory work in Europe, it is based on the advisor gaining an in-depth understanding of the individual farm through farm visits and telephone communication. Its advisory work is also financed through cost recovery.

Given the difficulties in the way of introducing FSD into governmental administration, and the urgency of the task, it would be extremely useful if the approach was spread as widely as possible among NGOs in the agricultural field. Each one would need to adapt it to its own particular needs and circumstances, but it would be of value to many. As time goes on more and more agricultural development is being channelled through NGOs: they become a more significant vessel for FSD work as a result.

3.3 Reorganising Agricultural Administration

There are four ways in which agricultural administration can be reorganised to facilitate FSD. These are: the introduction of cost recovery systems such that the services are constrained by the market; the decentralisation of resource allocation; the restructuring of extension, and of research activities. These four are complementary, but initiatives in any will have positive results, and will lead to initiatives in the others. All can be tried in restricted geographical areas or sectors until sufficient experience is gained. Indeed the costs of administrative reorganisation are usually so great that substantial reorganisation should not take place without experimentation. The reorganisations concentrate on both the demand for and supply of farming system development work.

3.3.1 Paying for Services

Many developed country agricultural services are already provided on a cost recovery basis. I have already described the cost recovery involved in Norway's farmer-managed research programme. Cost sharing systems are widespread in the US (Axinn, 1989: 96 ff), and the UK's Agricultural Development and Advisory Service (ADAS) has recently introduced payment for previously free services. The effect of this change has undoubtedly been negative in orienting the attention of advisors away from the smaller farms, but positive in stimulating a much higher quality of advice, tailored precisely to the needs of the client. The organisation has become more dynamic too.

It would normally be remarked that small farmers cannot afford to pay for agricultural services. However, it can be countered that many poor governments can no longer afford it either, and that in practice poor rural communities often do already pay part of the costs of extension or research. They may provide housing, food, and transport for extensionists. In some cases they even pay a small salary; in others they make payments in kind. For research they provide land and sometimes labour. Cost-sharing is already on the agenda, at least in extension.

From an FSD perspective, the positive reason for encouraging cost recovery is that it gives the farmer the status of customer. If the service is not up to scratch, has little to offer it will not be paid for and will collapse. If it is useful it will be able to grow and provide more services.

The problem with existing, often informal forms of cost recovery is that they do not give the farmers much power. A formal system, in which the salaries of extensionists and recurrent costs for running the service were at least partly met by their customers the latter would be able (and required) to direct the attention of the extensionists to problems they wanted solving and ways in which they wanted those problems solved.

An Oxfam-funded project at Kebkabiya in north Darfur, Sudan, provides a good partial example of this approach. "Current developments...include the development of seedbanks for millet in 12 villages. These are managed by local committees which are becoming the base for the development of other activities. Rotating funds for managing the supply of improved horticultural seeds for use in the wadis and on basement soils are established and this year are becoming the basis for the village extension service. The extension workers are to be paid by Oxfam in the first year (which also pays two thirds of the cost of a donkey for travel) but thereafter their salaries of £S 200 a month should be the responsibility of the management committee and be paid out of a levy on transactions in the revolving fund." (Curtis and Scoones, 1990).

In 1991 drought has slowed the project down. However, management committees have been collecting repayments and funds have been accumulating in the villages. The management of these funds remains an issue which the project has not fully resolved as it attempts to hand project management over to the management committees. It seems as though it would not be a problem to fund extension workers' salaries from the revolving funds.

Payment for extension by itself does not necessarily mean that extensionists will have something useful to extend. Indeed it will be easier to get farmers to pay if there is already a stream of useful information emerging from an organisation. Thus a marriage between the Oxfam extension approach and the adaptive research/farm system development approach of the Jebel Marra Rural Development Project would seem to offer much.

The Jebel Marra Rural Development Project (JMRDP) has developed a close relationship between its adaptive research and extension sections, and the generation of appropriate knowledge has been fed by a monitoring and evaluation service which has enabled scientists and extension managers to understand the farm systems and produce innovations tailored to at least some of them, if not the poorest. The project aims to develop and offer a choice of technologies in each recommendation domain. Thus in the field of animal drawn equipment both donkey and camel implements of different designs are offered to farmers. As a result of greatly increased camel, horse and cattle prices following the droughts of the early-mid 1980s donkey traction has become popular, and the project has responded to this, and indeed led the way in development of a new plough-weeder. Promising sorghum and millet seeds have been identified, and will be multiplied by contract farmers for widespread availability in 1991.

The project's extension service is organised in a classical top-down way though with the addition of annual village level "orientation meetings" to get feedback from the farmers. It has recruited agents from outside the area with the necessary formal qualifications, but these are apparently not reaching the 16% of households headed by women.

In 1990 it was proposed that the JMRDP include Kebkabbiya District in its purview. The outcome of this "marriage" is not known.

3.3.2 Decentralised Resource Allocation

As an interim measure, until farmers are able to finance their own extension and research services it should be possible to allocate a proportion of existing agricultural budgets to local farmers' organisations to spend on the services they want. They could then contribute to the funding of research which they had requested or which interested them and to the sort of extension or advisory system they found useful.

This would of course mean that they would be free to make contracts with extensionists/researchers outside the ministry structure. They might also decide to spend their money to train their own advisors/researchers whom they would then pay.

Under a structural adjustment programme this kind of resource decentralisation could be progressive, starting with a small but significant level of decentralisation, progressing, at least with organisations able to use the funds sensibly, to more substantial decentralisation.

The kinds of organisation to which such resources could be decentralised would vary from one country to another, and in some might not exist at all. They could be informal organisations based, for example, around farmer managed irrigation systems, or some other common resource: however, the danger here would be that the conversion to formality which budget decentralisation would require might kill the effective informality of the original. They could be a range of formal organisations: peasant association in Ethiopia, local resistance council in Uganda, ward committee in Zimbabwe (population c. 6000; number of farm households c. 7-800), co-operative societies etc.

Special local committees could also serve the purpose, where no existing organisation is adequate.

Resources need not be in the form of cash, but could be tokens or entitlements which would permit the purchasing of a limited range of services in the agricultural field. The range should be wide enough to permit private or co-operative sector advisory services to compete with the state. This will enable the now large pool of agricultural graduates who do not find employment with the state to use their expensively learned skills without being a drain on state resources. It might also encourage farmers to send their sons and daughters for agricultural rather than any other academic qualifications.

The chief problem with any cost recovery scheme is to protect the small farmer, who is less able to pay (though proportionately s/he may not have to buy very much), and less able to compete with the big farmer who can offer the extensionist more money and a bigger potential impact. This can be done by:

- allocating resources (tokens etc) on the basis of farm households rather than farm size;
- enshrining small farmers development in the objectives and reward systems of the agency;
- maintaining separate agencies for small and big farmers.

3.3.3 Structural Change in Extension Services

From the foregoing it could be suggested that a vital pre-condition for the widespread adoption of FSD is the availability of highly skilled agricultural personnel at local level. Currently most ministries post agricultural graduates and diplomates down to district level; some post diplomates to sub-district level. One of the reasons that FSD may be adopted more readily by NGOs is their ability to employ graduates at sub-district level. This should also be a target for ministries.

Since living and working conditions are in some (stereotyped) ways more difficult in the countryside than in the town it may be difficult to persuade agricultural graduates to work at this level. In the longer term agricultural ministries and colleges could give preference to farmers' sons and daughters in their recruitment. They would probably find it easier to live at sub-district level. In the shorter term it may be necessary to enable district level graduates to be more mobile, so that they can effectively work at sub-district level without committing their families to village life.

Incentive structures need reviewing. This is particularly complicated where civil service pay is depressed since compensation for low pay is sought in all kinds of allowances (travelling, meeting, night, day) and even bonuses, which bias activity in favour of activities accompanied by allowances, and which may not be equally available at different levels in the hierarchy. Under such circumstances special field living allowances could compensate for allowances unavailable in practice in the field.

Ideally the officer working at sub-district level needs to be part of a district level team, rather than supervised by officers at the district. A very shallow hierarchy, stressing teamwork among equals with varied specialist training, is the organisational structure which corresponds most closely to the FSD approach.

In fact some underdeveloped agricultural ministries already have these characteristics, since they have been unable to recruit the staff to fill a long chain of command. In particular, where employment of local extensionists has been passed to local government, while the civil service continues to employ diplomates and graduates (the situation in Sudan, for example), local government has not had the resources to employ numbers of lower level extensionists. The degree of structural change required in such a system may not be great: it would rather be a matter of changing work allocation, procedures and training in FSD.

The more difficult situation arises in a more developed agricultural ministry with a long chain of command and the classical bureaucratic features of fragmentation and specialisation. In such a system, sub-district level work is seen as belonging to local extensionists, while diplomates and graduates supervise. The basic task confronting an FSD approach is then to re-orient the supervisors, help them to build up teams of FSD-oriented graduates and diplomates around them, and train lower level extension cadres in the simpler aspects of FSD work. In the context of an FSD approach the usefulness of these lower level cadres is questionable, however.

Once confident in the competence of their teams, supervisors need to allow team members a greater degree of autonomy in decision-making than tends to be the case in the conventional extension approach. This autonomy implies a learning process: managers need to monitor and help their team members learn from experience as well as analysis. Once goals, objectives and a work programme has been agreed between a manager and a team member, the team member is expected to get on with the job. S/he must have the resources to do so: these should thus either be decentralised within the ministry, or, as discussed above, be lodged with the farmer.

The problem of availability of good seed varieties illustrates this point. Many ministries control very closely the release and multiplication of improved seeds. It is undoubtedly correct to regulate and monitor this important activity. However, the degree of regulation through licensing often means that an extensionist, having identified seed availability as a constraint, does not have the means to reduce that constraint locally, when in fact this could be done. FAO has advocated a model of decentralised seed multiplication by small contract farmers. In Somalia an FAO-sponsored project discovered a workable approach through trial and error between 1984 and 1987. At first a 200 hectare seed farm at Afgoi was the focus of efforts to multiply Somtux maize seed. However, irrigation limited the area to 50 ha. Three medium sized farmers were approached to grow seed on contract to expand the supply. However, these turned out to be not good farmers and did not follow instructions to ensure a quality product. So a third approach was tried.

A village with a good farming reputation and a co-operative chief with authority was chosen and contracts were made with all the farmers to produce only the varieties of maize determined by the Ministry. In the first year there were 50 contracts on a total of 50 hectares. Farmers were paid a 20-30% premium, and produced 50 tons. In the second year these figures were doubled. By 1990 the original village wanted to form a seed production company and were seeking a bank loan. (FAO, 1989)

Clearly autonomy and decentralised resource management is not enough to facilitate such developments in the farming system. It must be backed up with the relevant skills. These can be "bought in" from more central authorities, or indeed from the private sector.

3.3.4 Managing the Research-Extension Link

The chief difficulty which research services have had in introducing a farming systems perspective has been to institutionalise the process of on-farm testing and dissemination. It is well known that relations between research and extension tend to be weak. Research organisations seeking to engage in on-farm research can adopt one of three approaches: they can do it themselves; they can contract with the extension services directly; they can set up a co-ordination mechanism.

Where they do it themselves (and the extension services may also be doing the same) duplication and competition may be the result. However, this may also be healthy if there is just enough co-ordination to prevent gross duplication. The Ethiopian Institute of Agricultural Research has gone as far as setting up its own extension service to do on-farm testing and dissemination of results in a farming systems perspective. This may be a typical end result of seriously promoting FSR without a comparable re-orientation in the extension services.

In Tanzania, under a small research project investigating Resource Efficient Farming Methods (REFM) the research organisation (initially the Tanzanian Agricultural Research Organisation, now reintegrated into the Ministry of Agriculture in a restructured Research and Training Department) contracted directly with local extension departments in a number of districts to carry out experiments in low external input farming systems. These were mainly to investigate the use of farmyard manure in different forms and combinations with inorganic fertilisers, the use of leucena spp, and crop rotations. The on-farm experiments proved remarkably effective on a local level. However, lack of funds had brought REFM to a standstill in 1991.

At Mlowa Barabarani, a village 40 km from Dodoma, the REFM demonstration site had been taken over by Global 2000 when I visited in April 1991. Under REFM the local extensionist had been demonstrating the value of FYM for continuous cultivation of sorghum over a number of years. Yields had reached 15 bags (of 100 kg) of sorghum per acre with FYM, compared to an expectation of 4 bags or so without. Last year the same sorghum HYV yielded 12 bags with the application of Global 2000's package of 50 kg urea and 25 kg TSP. This year the same formula was expected to produce more than 15 bags due to better weather. On contact farmers' fields both the sorghum and millet crops had been fertilised and responded well.

Farmers were universally enthusiastic about FYM, and the local extensionist claimed that 80% of the village's farmers now used it. On the other hand they were sceptical about inorganic fertilisers, believing that they might exhaust the soil. Farmers also appreciated the moisture conserving properties of FYM given their rainfall pattern (average c. 400 mm per year, range 300 -700 mm). The extensionists were flexible, and thought that probably the combination between FYM and inorganic fertilisers was best. They also thought that crop rotation with a legume would be important to restore the soil after a couple of years of fertiliser use.

Certainly the economics of subsidised inorganic fertilisers is favourable to the farmer: the Global 2000 "package" of seeds and fertilisers costs less than £7 per acre, and is available on credit, an amount which can be repaid with a small sale of produce. But for those who have FYM and family labour (the majority in Mlowa and the Dodoma area) this source of fertility is even cheaper, and therefore preferred by farmers. The constraint on FYM use is transport from the kraal (boma) to the field. Few farmers own ox-carts. Many headload it. Not surprisingly fields further away from the boma receive less manure.

Some farmers experimented with FYM on part of a field and inorganic fertiliser on another, and combinations of the two: the result, perhaps, of years of encouragement through on-farm trials.

The labour required for FYM operations is needed during the dry season - when other demands of family labour are minimal. By comparison the Global 2000 package increases labour input at planting, as 25 kg of urea is mixed with the TSP as a basal application with the seed. This means that a third person is required to apply the fertiliser in addition to the two normally required to dig a hole and sow the seeds.

The early maturing (60 days) millet (Serere, from Uganda) distributed by Global 2000 proved to be less responsive to fertiliser than the local (100 day plus) varieties. However, anything that increases farmers' choice is probably a good thing.

The Global 2000 project had been able to take over the on-farm trials simply because it had the resources to offer the extension services. This illustrates the problems faced by a poverty-stricken ministry in developing a consistent approach to technology improvement. Undoubtedly an FSD approach would have helped extensionists be clear about the relative values to different types of farmers of the innovations tested on farm.

The impact of REFM's results and approach have been much wider, however. In the current reorganisation of agricultural research in Tanzania many research projects will compare and combine FYM and inorganic fertiliser use as one of many research foci with the objective of promoting low input technologies, minimising environmental pollution, and developing sustainable soil and water management practices; and each zone of the country is to have a research/extension liaison officer and an FSR team to manage on-farm trials. This is a substantial reorganisation, which may however, come up against the same problem as the in Ethiopia: the lack of orientation of the extension services to a farming systems perspective.

In Zimbabwe a Committee for On-farm Research and Experimentation has been set up to co-ordinate between research and extension services within the Ministry of Agriculture. However, the actual work of this unit seems to have been limited to a few FSR projects in specific locations. It has not been generalised, and the perception of Agritex, the Ministry of Agriculture's powerful extension organisation, is that little of use has emerged from farming systems work.

4. MEETING THE INSTITUTIONAL CHALLENGE: COMBINING ADMINISTRATIVE REFORM WITH STRUCTURAL CHANGE

As demonstrated by the Nigerian example quoted above, much can be done through innovative training to inspire change in practices. It is notable that this success was in the context of local government establishments which had relatively recently assumed new and increased powers and resources, but which had ill-defined procedures for utilising these. In fact Agriculture Departments of local governments were desperately short of ideas for useful projects and farm innovations: the FSD approach would fit there admirably.

In other circumstances training for FSD might be a waste of resources. Where there is a highly organised T & V system, or other top-down extension process, it is unlikely that FSD will be accepted by top management.

Conducive circumstances for FSD might be:

* geographical areas where top-down approaches have clearly failed
* extension systems with well trained and under-used generalists at sub-district level
* the absence of a well organised T & V approach
* sustainable agriculture programmes, which take indigenous technical knowledge seriously
* NGOs seeking to develop participatory agricultural extension services
* good research-extension links.

It was argued above that decentralised decision making was the institutional key to a sustained FSD approach. Clearly, FSD work should be located as a matter of priority, at this early stage of its career, in systems which have already reformed themselves significantly in the direction of decentralisation. Systems which have permitted local autonomy in resource control and use, and an experimental approach to development should be sought out by FSD proponents to maximise its use.

Thus there are opportunities for FSD work in both reformed and unreformed situations. However, as with IRD programmes, organisational factors should always be featured strongly in the design of any FSD activity.

At this innovatory stage the need is to prove the utility of the concept: for this favourable organisational circumstances should be identified, where an experimental and different approach to agricultural development can take root.

REFERENCES

G. Axinn (1988), Guide on alternative extension approaches, FAO, Rome

R. Chambers (1974), Managing Rural Development, Uppsala 1974

J. Cohen (1979), The administration of integrated rural development projects, Harvard Institute for International Development Discussion Paper No 79

D. Conyers et al (1987), The role of integrated rural development projects in developing local institutional capacity, Iowa State University, Studies in Technology and Social Change, No 2

J. Copestake (1990), The scope for collaboration between government and private voluntary organisations in agricultural technology development, Overseas Development Institute, Agricultural Administration Network Paper 20

D. Curtis and I. Scoones (1990), Strengthening Natural Resource Planning Capability in Darfur, Report for the Agricultural Planning Unit, El Fasher, Darfur

J. Field (1983), "Development at the grassroots: the organisational imperative" in D. McLaren (ed) Nutrition in the Community, Wiley

FAO (1990), Farming Systems Development: guidelines for the conduct of a training course, Rome

FAO (1989), Somalia: Seed Production Strategy, Rome

A. Fowler (1990), Non-governmental organisations in Africa: achieving comparative advantage in relief and micro-development, Institute of Development Studies, University of Sussex, Discussion Paper 249

I. Goldman et al (1989), "Facilitating sustainable rural development: an experience from Zambia", Journal of International Development Vol 1 No 2 pp 217 - 230

G. Honadle and L. Cooper (1989), Beyond Co-ordination and Control: an inter-organisational approach to structural adjustment, service delivery and natural resource management, in WORLD DEVELOPMENT Vol 17 No 10 pp 1531 - 1541

G. Honadle et al (1980), Integrated Rural Development: making it work ? Development Alternative Inc, for USAID, Washington

E. Morss and D. Gow (1981), Integrated Rural Development: nine critical implementation problems, Development Alternatives Inc, Washington

A. Shepherd (1991 forthcoming), Method and anti-method in project identification, Papers in the Administration of Development, University of Birmingham

FARMING SYSTEMS DEVELOPMENT - CAN IT BE INSTITUTIONALIZED?

by

N.R.C. Ranaweera[1]

1. INTRODUCTION

Farming systems has during the last 20 years taken the lead role in influencing the research agenda in a number of countries, particularly in the Asian region. With this long experience, however, there still exists grey areas which raises issues dealing with the real value of Farming Systems approach to overall agricultural development.

Many are the experiences where countries have institutionalized farming systems research into the research and the farming systems perspective into development programs. However, in the field, there still exists many researchers who prefer the commodity approach, who believe in direct extension activity for the spread of new technology.

Institutionalizing FSR confined itself to the research institutions within most of the countries, extension involvement though theoretically required, does not exist in most countries. They usually are an afterthought, which are entertained often to find that it is too late.

This particular discussion on the institutionalizing of FSD is therefore timely.

In this presentation, an attempt is made to raise issues as to why even after the institutionalizing process, there still exists a task of commitment on the part of certain countries and further why those that have institutionalized find the benefits marginal.

2. FARMING SYSTEMS RESEARCH (FSR) AND FARMING SYSTEMS DEVELOPMENT (FSD)

It is now over two decades that researchers began using the FSR or more the "Holistic" approach to be the underpinning base for agricultural research. While The concepts and approaches have been accepted and much headway made.

---

[1] Deputy Director of Agriculture and Head, Division of Agricultural Economics and Planning, Department of Agriculture, Peradeniya, Sri Lanka.

There still are the "doubting Thomases" who question the approach. Nevertheless, most agricultural research systems have formalized and absorbed into the research activities the FSR approach.

However, most activities have stopped there. The technological findings that were developed within this perspective could not be translated into farmers fields since most extension activities to date are still commodity based. So are the many supporting services such as input and credit supplies and marketing organizations.

The question is then asked does FSR have to be extended beyond into an FSD process. This essentially requires the researchers, extensionists and more important policy makers to consider the farming systems approach into the development process in the agriculture sector in a country.

Has this been achieved in any conceivable manner in the countries? The answer is no and the reasons are many. There are many issues that arise which if considered could provide directions to make FSD a basis for agricultural development.

3. ISSUES IN INSTITUTIONALIZING FSD

For any research and/or development organizations to institutionalize an activity, these must be adequate justification that can be provided to decision-makers. Up until very recently, agricultural development was considered in a unimodal sense, even though development activities are multisectoral activities. The main approach towards increasing farmers real incomes were through increases in yield and overall production. It can be shown that agricultural production has increased in a number of countries during the last decade but, whether the real incomes of farmers have increased in the same manner in these countries are unclear (Table 1). Most donor agencies viewed agricultural development through investments in research and extension (eg. T&V) - again on a crop basis. It is after almost a decade of such activities that donor agencies have now realized that what is required is a more "holistic" approach to agricultural development. It is within this background that some of the issues in incorporating FSD as an approach into the institutions dealing with agricultural development are discussed.

3.1 Appropriateness of Technology Development Process

The last decade has been full with significant claims of new technology with FSR an perspective. In reality though most of these technologies are commodity oriented and very few are systems oriented. In Asia, a careful examination of the technologies developed will indicate that component technologies developed are still single crop oriented and concentrated on changing crop sequences or introduction of new crops. Little (if any) consideration has been given to the interactions between the different components of a farm (viz: highland - lowland - animals - homegarden). The changes in the resource base too have not been considered particularly in relation to the limited resources of the farmers, namely land, labour and capital. Essentially technology development has been independent of the resource base of the farmers. Consequently, when recommendations are made to farmers, they have in the past tended to be only component oriented and not necessarily resource oriented.

This has led to difficulties in promoting such technologies. Extension programs have, as stated earlier, concentrated on the commodity/crop component not necessarily the system that existed in the farm. Commodity production promotion programmes took priority over increasing overall farm incomes through consideration of the farm-off, farm-non farm interactions. Therefore, from a systems point of view not only the component technology, but also the extension efforts were non-existent.

3.2 Involvement of Farmers in Technology Development and Subsequent Production Programs

The technology development process is ideally said to be interdisciplinary in nature - mostly from a scientific sense. Assuming the process to be primarily on-farm research based, the involvement of the farmer becomes critical. However, in reality, the involvement of farmers tend to be on an "observer" status. Not always are the farmers aware of the purpose of the experiments being undertaken on their fields. This leads to the situation where farmers are unable to appreciate adequately the findings of the research activity. When a production programme is then to be undertaken, farmers have essentially got to be reoriented or retrained, thus loosing valuable time in obtaining the results.

This situation need be corrected if maximum use is to be made of the technology being developed. Attempting to encourage farmer participation after the technology is developed, is not only time consuming but is also contrary to the very nature of the systems approach in the development process.

Recently, there have been People Participation Programmes (PPPs) in a number of countries. The purpose of these programs are mainly to organize farmers individually, collectively or cooperatively in production, marketing and utilization programmes. However, very little if any technology development is undertaken in these programs. In fact, research scientists have often not been involved in PPPs in the development of PPP's both farmers and extensionists can be involved collectively in planning and organizing the programs, it is an achievement for the researchers to provide technical backstopping to the PPPs is the ideal. PPPs should also gear to the socio-economic and cultural environment, in which case the acceptance of the technologies will be better.

3.3 Sensitizing Extension Workers to FSD

Extension programs in most countries are commodity oriented. This situation has been prevalent over a number of years. To change it will require a number of measures, the first of which is to sensitize the extension workers to the "whole farm" concept. They are probably the easiest to influence since they deal with the whole farm in their day to day activities. Rarely are training programs organized for the extension workers where he is required to understand the interactions as a consequence of changing the resource base of a farmer, due to a new technology. They must be informed of possible changes in marketing prices and demand for such commodities. Most extension activities are production oriented and hardly. Is there any effort to understand the market and input supply system. Any institutionalizing efforts of FSD should include this aspect.

3.4 Involvement of Support Services

Technology has been developed over a number of years on a FSR basis. However, the support services particularly credit, marketing and input supplies have been commodity oriented. All countries talk of credit for paddy or another commodity production programme, or supplying inputs such as fertilizer or chemicals for a single crop. Little attention is given for credit or other inputs to the rest of the farm of homegarden of other animal activities, which in almost all Asian countries is very prevalent. This is very important. Efforts must be made to sensitize credit and marketing agencies to the value.

For a successful FSD program, the issue is how well can the different support institutions perform their functions within this framework? It needs close interactions and coordinations.

3.5 Linking FSD Programs at the Community Level

Most development programs are macro in nature, usually planned and implemented form the centre. However, day to day implementation is at the community level, involving extension and other community level workers. Integrating FSD programs into community level activities is imperative for maximum impact of FSR and extension activities. Community level worker training to appreciate the systems approach is needed.

4. WHY HAS INSTITUTIONALIZING FSD/FSR BEING DIFFICULT?

As identified earlier, the systems approach has been in the "system" for over two decades. However, efforts are still being made to have FSR placed in a formalized institutional framework. Is this delay due solely to a lack of understanding or misunderstanding of the system, or is it due to the fact that National Agricultural Research Systems (NARS) see marginal value in it?

For over two decades these have been concentrated efforts to institutionalize the FSR/D approach in NARS. In the Asian regions, certain countries have created institutes to reflect this e.g. Thailand, Philippines, Nepal, India, Bangladesh. However, in all these countries, this aspect is reflected mostly in the FSR components only.

These appear to be a lack of appreciation for this approach is development aspects. Could this be due to the fact that NARS and its policy-makers consider the approach not much different from the traditional approaches in agricultural development which has borne fruit as reflected in the green revolution? It is because that it does not still reflect a bottom-up approach and is mostly confined to the research-extension officers whose perception is not necessarily shared by the policy makers? Is it due to the popularization of the approach being donor driven? This is reflected in a number of FSR/D programs usually being stopped once donor funding ceased. Examples of this has been in many countries in Asia.

5. CONCLUSION

This paper has attempted to raise certain issues that in my view rather basic to the entire question trying to institutionalized FSR/D within NARS. While there are examples of formalizing the approach in the research systems, the acceptance of it as a major policy measure in the implementation of agricultural development programs still remain weak.

What is required is an innovative and more appropriate approach to make practitioners and policy-makers aware and appreciate the approach.

Table 1. Production changes of selected agricultural crops 1980-89 (000T).

Year	INDIA Rice[a]	INDIA Wheat[c]	PHILIPPINES Paddy[a]	PHILIPPINES Sugarcane[b]	SRI LANKA Paddy[a]	THAILAND Paddy[a]	INDONESIA Paddy[a]
1980	80312	31564	7723	2273	2133	17368	29652
1981	79883	36460	8122	3208	2229	17774	32774
1982	70772	37833	7731	3588	2156	16879	33584
1983	90048	42502	7841	3123	2484	19549	35303
1984	87553	45148	8200	3790	2414	19905	38136
1985	95818	44069	9097	2293	2661	20264	39033
1986	90779	46885	8958	2071	3588	18868	39727
1987	85339	45576	8540	1839	2128	18483	40078
1988	106385	45096	8971	1883	2477	21263	41676
1989	106220	-	9651	-	2063	21000	44779

Source: a. IRRI 1990 World Rice Statistics
b. Philippine Statistical Yearbook, 1989
c. FAO Production Yearbook

FARMING SYSTEMS RESEARCH AND RURAL POVERTY: A POLITICAL ECONOMY PERSPECTIVE ON INSTITUTIONALIZATION

by

S. D. Biggs[1]

1. INTRODUCTION

In recent years there has been a growing interest by donors and international research and development (R&D) institutions in promoting various forms of farming systems research (FSR) to reduce rural poverty. Recently the scope of FSR has been extended to encompass a broad range of agricultural extension and overall agricultural and rural development planning and implementation activities.[2]

With this increased interest in FSR there has been an expansion in programs to promote the application of FSR principles and methods.

However, reviews of FSR programs have shown that the implementation of FSR is an extremely complex task. The actual outcomes of projects are often quite different from the projections of planners. Although the participation of resource-poor farmers in the research process, and the promotion of 'feed-back' to influence the priority setting of experiment stations, have been two central objectives of FSR programs, the evidence from a recent nine-country study shows that the performance of most projects on these two central criteria, has been very disappointing.[3]

---

[1] School of Development Studies, University of East Anglia, Norwich

[2] For a review of approaches see Merrill-Sands (1986) and Chambers and Jiggins (1987, 1987a). The expansion of FSR to include extension concerns is reflected by the Florida University's Farming Systems Research and Extension (FSRE) programme. An even broader approach to include all aspects of agricultural, and rural planning is suggested by FAO in their Farming Systems Development (FSD) concepts (Friedrich and Hall, 1990; Dixon, 1990; FAO, 1991).

[3] A recent international study of on-farm, client-oriented research (OFCOR) by ISNAR of about forty FSR projects in nine different countries found, amongst other things, that resource-poor farmer participation in research was very low and only occasionally did research stations respond to 'feed-back' by changing their priorities

In this paper it is argued that two of the reasons for these 'unexpected' outcomes are that:

1. Political and institutional analysis, in the context of FSR, has been neglected in the past.

2. There has been over emphasis on the transfer of new techniques from international centres to developing countries (and from national centres to regional or sub-regional programmes), and not enough attention to understanding how techniques and approaches arise from multiple institutional sources over time and space.[4]

In order to illustrate these points, we shall look at some examples and draw out some conclusions for the implementation of FSR. However before proceeding it is useful to outline a broad framework for analysis.

2. A DYNAMIC FRAMEWORK FOR RURAL DEVELOPMENT

FSR can be thought of as part of a continuous process of research, taking place in villages and households. It includes all the selection, experimentation and other 'informal' research activities of farmers, village artisans etc and the village level activities of natural and social scientists in the 'formal' research system.

Alongside these village level activities are research station activities such as controlled experiments and literature reviews by social and natural scientists. New ideas come from both village-level research and from researchers who spend most of their time in universities and research institutions. This is represented in Figure 1 by the arrows between village-level research and experiment stations. Research is influenced by the political, institutional and economic environment in which it takes place. This is reflected is by the pattern of funds allocated to research. Institutional policies which govern pay scales, status and other rewards in the formal research system are equally important aspects of science policy. These relationships are indicated on the diagram by arrows going from research categories to the political, institutional and economic context.

(Merrill-Sands et al, 1989). The ISNAR study looked at the reasons for different levels of performance for the following seven functions: applications of a farmer-oriented, problem-solving approach, application of an interdisciplinary systems perspective, characterization of agro-climatic and socio-economic groups, adaptive research, farmer participation, feed-back to research priority setting, collaboration with extension and development agencies.

[4] For a literature review which supports this suggestion see Biggs and Farrington (1991). The details of a multiple source of innovation model are given in Biggs (1990).

Figure 1: **A Political Economy Framework for Farming Systems Research**

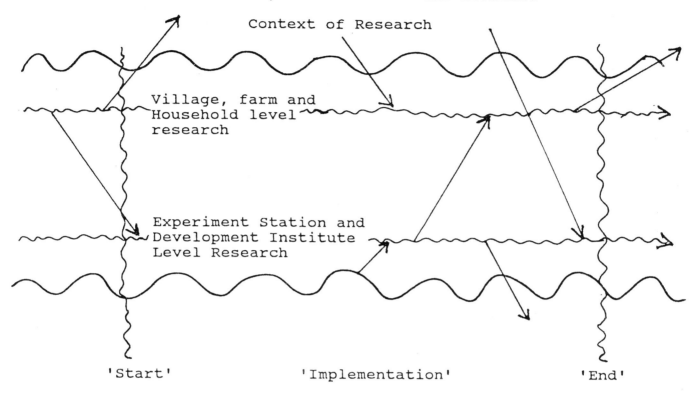

Note

a) The 'start' and 'end' vertical wavy lines represent the beginning and end of a research project.
b) The large horizontal wavy lines represent the boundaries between the external political, institutional and economic context of research and the research organisation. The lines are wavy to indicate that these boundaries are often hard to define and 'permeable'.
c) The small horizontal wavy lines represent the on-going programmes of the research stations and village-level research.
d) The arrows represent the way that different factors affect the research programmes and the way research can influence the context in which it takes place. The arrows illustrate that there is no single one-way causation in the relationship between research and the political context.

(Based on Ashford and Biggs, 1991)

Research is a dynamic process which constantly interacts with society, which is itself continuously in flux. Researchers are always involved, explicitly or implicitly, in directing the path of science and influencing the distribution of benefits of technical and institutional change. The wavy lines at either end of the research lines demonstrate how it is difficult to define the beginning and end of a project in a clear-cut way. An example of this is the difficulty within research assessment exercises to give adequate recognition to past research investments. By the same token, some important research 'findings' have to wait until the institutional and political context becomes conducive to recognition of the work.

It is important to see this analysis as a general framework which can be applied to any situation. The political economy perspective does not see planning as necessarily a hierarchical exercise. It takes a particular situation and looks at what forces - whether they be local, national or international - caused changes to take place. For example, the development of some new approaches in international research institutions may have been influenced directly by the work of practitioners in developing countries. On the other hand, in other situations the causation may have been the other way round.

Another feature of this political economy framework is that it stresses the importance of using language and concepts carefully. It recognises that these are never 'neutral'. This is particularly important for a subject such as FSR and rural development. What may be called a 'success' when one set of criteria are used may be a 'failure' when another set are used. In the present context, the term "Institutionalize farming systems development" needs to be used with considerable caution because:

a) it might imply that there is agreement on the definition of FSR, and
b) it can create an impression that the ideas or components of the package did not exist before the "Institutionalization" process began. There are always existing formal or informal institutions which conduct some or all of the functions designated in any new institutional package. There is never a 'clean slate'. Interventions can strengthen some types of local institution and weaken others. In addition, the actors involved in bringing about different types of institutional change are never disinterested analysts. They all have institutional and personal interests which they pursue in different ways.[5]

In the next section some case studies of FSR practice are used to illustrate why it is important to place 'successful' FSR practice in its political and institutional context, in order to understand what gives rise to desirable processes of institutional change.

---

[5] For a discussion of the use of language and the role of different institutional actors in agricultural and rural development projects see Clay and Schaffer (1984) and Wood (1985).

3. EXAMPLES OF FARMING SYSTEMS RESEARCH PRACTICE

3.1 Development and Use of Rapid and Informal Surveys

A major feature of FSR is the use of informal surveys, reconnoitre surveys, exploratory surveys and rapid rural appraisals. An example of the effective use of surveys was Bangladesh in the early 1970s, soon after independence, where a range of quick, unstructured surveys were initiated. In the Ministry of Agriculture, under the auspices of the Agricultural Research Council, interdisciplinary teams of social and natural scientists from research stations, NGOs, and universities conducted quick surveys in all the districts of Bangladesh, to learn about changes in agricultural technology. The workshop that followed the rapid surveys in Bangladesh, and other parts of Asia, discussed major issues of future agricultural research policy in Bangladesh (BARC 1975). Under the auspices of the Ministry of Local Government and Rural Development, rural surveys were also initiated to learn from the great variety of government, NGO and private sector rural development programmes in many parts of Bangladesh. All the major universities and social science research institutions were involved in organising teams of staff and students to visit, learn from, and write about these diverse institutional initiatives in rural change.[6] One of the organisers of these surveys was Dr Md Yunus who,

---

[6] The survey work was organised by the Central Evaluation Committee of the Ministry of Cooperatives, Local Government and Rural Development. The purpose of the work was explained in the foreword to one of the reports from the Economics Department at Chittagong University. It says:

'During the last two years a new phenomenon has been taking shape in Bangladesh - instead of waiting for the national Government to design plans, send out directives and funds, the local leadership began taking confident steps on their own, to design their own plans and implement them to make their areas self-reliant. Leadership came from diverse quarters. In some cases imaginative, resourceful youths provided this leadership, while in others political leaders came up with the quality of leadership people of Bangladesh almost forgot to expect from them. In still others it came, again quite unexpectedly, from the district administrations. Implicitly or explicitly they all wanted to transform the beggars 'extending arms into confident productive hands' of a hard-working nation.

The characteristics of these local programs are as diverse as their location, area of coverage and leadership pattern. Some covered entire districts involving all branches of administrative machinery and other institutions, some concentrated in very small areas experimenting with new institutional designs for rural development. In terms of organisation and efficiency they varied widely. Not all of them could, however, retain their initial enthusiasm. But the successes are striking.

Realising the importance and significance of these experiences dispersed in time and space, the Ministry of Local Government and Rural Development decided to evaluate and document these experiences, as their sponsors grope through an uncharted course. A Central Evaluation Committee was formed by the Ministry for

at the time, was experimenting, with support from a local Agricultural Bank, to develop a new institution to provide rural credit to landless women and men. During the 1970s and 1980s the Grameen Bank evolved, and is now a major bank in its own right, providing viable credit facilities to some of the poorest people in Bangladesh (Hossain, 1988).

Bangladesh provides another example of an effective rapid survey. The report from a quick survey by Ahmed (1975) described the way handpumps (originally introduced to provide drinking water) were being used for irrigation purposes and were spreading rapidly. Richer farmers were buying the pumps and charging very high rents to landless labourers and small farmers who could not afford them. This survey report was one of the important documents that led to an innovative new Ministry of Rural Development/UNICEF programme, providing credit to poor people for the purchase of pumps, and for NGOs to initiate R&D to reduce the drudgery of the work and improve the economic efficiency of the pumps.[7]

There are many other examples of rapid rural surveys which have been an integral part of a wider economic and political reality. Although many of the survey methods and institutional arrangements were innovative, it is important to note that the wide variety of actors involved did not set about developing a common set of new methods which were to be replicated and widely promoted. The methods were developed by each group as a viable response to what was seen by them as a local research and analysis problem.

As an example of the development of new organisations, the new Central Evaluation Committee in the Ministry of Cooperatives, Local Government and Rural

---

this purpose in May 1975 with Mr Mahbub Alam Chashi, the then Vice-Chairman of the Bangladesh Academy for Rural Development, as the Chairman.

This report is an outcome of the work commissioned by the Central Evaluation Committee with the financial support from the Ford Foundation. We hope this effort will help us all, in some way, in appreciating our potentialities for overcoming our problems and limitations'(Yunus, 1977).

Many universities and research institutions were involved in the quick rural surveys. The Economics Department called their papers 'A series of reports on locally sponsored Development Programming' and at least 28 were produced by 1977. The Economics Department at Rajshali University called their reports 'Rushed Reports'. Other institutions involved were The Bangladesh Agricultural University and the Bangladesh Development Institute. They had a different name for their series.

[7] For details of the way policy, institutions and R&D activities on manual irrigation methods and techniques interacted during the 1970s and 1980s see Biggs, Edwards and Griffith (1976), Biggs and Griffith (1987) and Orr et al. (1991).

Development was not created as an experiment with the view that, if it was viable, it would be transferred to other ministries. It was designed for the specific circumstances.

3.2 Influencing Agricultural Research Policy

One of the most important objectives of all FSR projects is to collect and analyze information which can be used to influence the direction of agricultural research policy. This can be done either by initial diagnostic surveys or by 'feed-back' as the work proceeds. However experience over a wide range of FSR and operations research projects has shown that feed-back has had little impact on research policy[8]. In the light of this bleak picture it is worth looking at some situations where diagnosis research and feed-back actually occurred. These cases are taken from India.

In Uttar Pradesh in the late 1970s, maize research priorities and breeding strategies changed in a large regional crop improvement programme, in response to on-farm research. Research increased on the improvement of maize varieties grown under stress conditions. These were the conditions under which the great majority of maize farmers in Uttar Pradesh operated. The programme also shifted research resources towards the problems of resource-poor farmers. Significantly, from an institutional point of view, the head of the maize-breeding programme and senior research station scientists were involved in all of the on-farm research, and wrote the regular annual on-farm research reports. This work included implications for changes in research policy. In effect they were making recommendations to themselves. In retrospect another significant feature of the case was that the University was closed at the start of this innovative work and scientists, who normally would not have had time to be away, were able to conduct research in the rural areas.[9]

Faizabad in Uttar Pradesh provides another case in which plant breeding priorities changed in response to the local diagnosis of technical problems of farmers growing rice under complex, diverse, deep-water conditions. In this case an eminent breeder developed, over a number of years, a panel of farmers who he interacted with continuously, by visiting their villages. Significantly, the plant breeder was equally innovative in his research station activities, where he utilised

---

[8] For examples see a summary of the ISNAR studies of client orientated on-farm research (Merrill-Sands, et al., 1989a) and Desai and Patel's (1986) assessment of the experiences of Operations Research Projects in India.

[9] See the annual reports by the maize improvement group at the GB Pant Agricultural University (e.g. Agrawal et al, 1978) and Biggs (1983). The annual reports which each year drew out implications for research policy, extension advice and for changes in on-farm research are classic examples of how to conduct and report high quality research in a timely way.

pits left from highway earthworks in order to create conditions for rice crossing and selection activities. The costs of these pits was a small fraction of what a conventional research station would invest in deep-water rice breeding facilities[10].

In both of these cases it is found that on-farm research was totally integrated with the research station priority setting because it was under the same management. In addition, in both programs the scientists were innovative in creating and developing their own techniques and analysis that proved relevant to their own institutional context. Both programmes had made a commitment to work on the problems of poorer rural households. Ideas and methods from outside were reviewed, screened, 'depackaged', and specific components used where appropriate. In the same manner international visitors to these two situations looked, learnt and selectively used ideas in developing their own advocacy and approaches to FSR.

Finally, a case from Andhra Pradesh provides a different type of institutional example, where 'feed-back' from on-farm research was used effectively to influence the priorities of scientists in a university. In this case the Vice-Chancellor of the Agricultural University took personal responsibility for reading on-farm research reports from a great variety of sources. He also talked regularly with researchers who worked in the field. In this case the Vice-Chancellor gave his authority and a major commitment of his time to trying to ensure that university research responded to local needs. Significantly the Vice-Chancellor, in a report on research monitoring and evaluation (M&E) in Indian Agricultural Universities, had earlier noted that past M&E had been minimal[11].

3.3 Introduction of Anthropologists

Another major objective of many FSR programs has been to include anthropologists and sociologists in research programmes. Perhaps one of the best known examples of this is the work of anthropologists in the programmes of the International Potato Institute (CIP), Lima. For many years they have advocated approaches associated with farmer participation, farmer experimentation and quick informal surveys. In a recent account of the involvement of anthropologists in the early years of the CIP social science programme there are three significant factors to note. The head of the programme:

---

[10] For details of the programme see Maurya (1986) Maurya, Bottrall and Farrington (1988).

[11] See the article on M&E by Appa Rao (1986). The report by Vishnumurthy (1987) entitled 'Feed-back from Operation Research projects and Reorientation of Research Programmes' clearly shows how a systematic approach can be taken to bring about changes in the priorities and work of a broad range of technology R&D programmes.

a. allocated a major part of the budget to employing anthropologists rather than economists, statisticians etc.
b. arranged transport for his staff when the head of the institute vetoed his requests for travel funds to conduct on-farm research.
c. encouraged his staff to develop their own methods and approaches when working with engineers and other natural scientists. These experiences were written up as detailed case studies.

In subsequent years funds were provided for on-farm research, and it has become a major part of CIP activities. Significantly, we find that the institution followed through with its policy commitment to the inclusion of anthropologists, and created a context for the social science division which promoted:

a. anthropology
b. the allocation of time to writing case studies which cover theory and practice in interdisciplinary research
c. advocacy of the CIP approach through practitioner examples - which included graphic reference to the continuous problems and conflicts of conducting interdisciplinary research - rather than by 'how to do it' manuals of methods and procedures[12].

3.4 Strengthening Linkages between Research and Extension Organisations

Perhaps one of the most challenging of the institutional questions facing FSR practitioners is how to strengthen the institutions which can provide useful linkages between extension and research organisations. Two cases will be used to illustrate

a. the importance of looking at cases where practitioners have been creative in different ways, in developing institutions to address this issue, and
b. how an analysis of the policy environment was important to understanding the rationale for the development of viable institutions.

3.4.1 Networks and Newsletters

In the early 1970s a number of non-governmental organisations (NGOs) formed an association called the Agricultural Development Agencies in Bangladesh (ADAB). One of the major functions of the organisation was to produce a newsletter containing information on agricultural and rural development. The

---

[12] For an account of the early days at CIP see Horton and Prain (1989). The diversity of approaches used in the early years, details of the case studies and lessons from social science research in CIP are given in Horton (1984).

editorial policy of the newsletter was biased towards agricultural and rural development subjects specifically concerned with reducing rural poverty. The newsletter carried material written by a large range of people in government research, education and extension agencies, NGOs, aid agencies and some people writing in a private capacity. The newsletter has a very wide distribution in Bangladesh and even an international readership. It has been an important communications network, providing high quality information between a large range of actors involved in rural development. It is unlikely that there are major issues of rural development policy and practice (including all aspects of FSR) that have not been addressed in the ADAB News.

What is significant in this case is that the need for a network to supply rural development information was recognised 20 years ago. The constitution of ADAB was drawn up so that it had a poverty focus and was not vulnerable to closure by government; ADAB had support not only from NGOs but also from a wide range of people in government and aid organisations. ADAB personnel continuously put time and effort into finding and editing relevant material and getting funds for the ADAB News. Finally it is critical to note that ADAB and the ADAB News has had its critics and its internal fights over such things as editorial policy. There have also been pressures at different times from government agencies for ADAB to behave in some ways but not others. ADAB, like any organisation which wishes to effect changes in society, expects pressures from different interest groups.

These institutional issues will continue to be of central importance to ADAB. The relative importance of arguments for or against alternative courses of action cannot be determined a priori and will depend on the political and institutional context at the time. Information systems analysts and those interested in promoting networks directed at poverty issues in third world countries could learn a great deal from a study of the creation and survival of the ADAB News and has survived over such a long period[13].

3.4.2 Regional Research and Extension Centres

In the Himalayan Hills, two agricultural extension centres were established in the 1960s to provide technology and extension advice to farmers in two zones of Nepal. The extension approach started with farmers being brought into the centre for training. It was soon found that this approach failed because staff at the centre did not know enough about the ever-changing production conditions of farmers to be able to offer suitable advice. The extension efforts shifted to the villages, and increasingly the centres found that they did not have suitable new

---

[13] In India a major problem is how to promote communications between different ministries, research organisations etc. Innovative ways to address these issues formed a major part of the work of a recent national workshop on Management of Research on Rainfed Regions (Gupta, 1988).

technology to transfer to farmers. This led to the development in the centres of a research capability. In 1984, in order to reduce the pressures on senior staff to remain at the research centre, the director of the Lumle Agricultural Centre, supported the suggestion for a new institution called the combined trek[14]. This became a regular trek for several days by senior and junior staff through the regions being served by the centre. The director generally led these treks. Throughout the trek there were many individual and group meetings with farmers. At the end of the trek, the implications of the discussions for the research station priorities were drawn up. Over the years there have been major shifts of emphasis in the way research is organised at the Centre, and in the technical research priorities of the Centre's work. As a result of the relevance and success of the group treks in Lumle, similar treks have been initiated in the other agricultural centre and in the national agricultural research programmes.

However, what is significant from an institutional perspective is that it was the director of the Lumle Agricultural Centre who was convinced of the need for and usefulness of the group treks, and introduced a policy to support them. Not only did the director personally make a commitment to the trek but he ensured that senior staff regularly participated and that funds and other resources were made available. Implementing this policy was not easy. There were (and still are) arguments by senior scientists that they were wasting their time by being in the villages, that the treks require considerable resources that could be better allocated to more productive activities. In addition, on some occasions the director did not respond to requests to attend meetings with his superiors in Kathmandu because he placed higher priority on participating in the trek. It is unlikely that similar types of group treks will be effectively implemented on a regular basis in the national programme, without sustained changes in national research policy to encourage senior staff to spend time in the field, and unless appropriate resources to be provided. This would mean a major change in overall national research policy which has for several years been based on a centralised system, with central research stations for major crops and associated national coordinated programmes.

This illustrates that even for the transfer of an institution, such as the group trek, there are major policy, economic and institutional issues that have to be addressed. Without this type of assessment the new institutional model might be irrelevant or even counter-productive if it is 'badly' implemented.

4. DISCUSSION AND LESSONS FROM RURAL RESEARCH PRACTITIONERS

4.1 Placing FSR in a broad Framework of Science and Technology Promotion

---

[14] For an account of the early days of the combined trek see Kayastha, Mathema and Rood (1989). Details of the present priorities and programmes in Lumle are given in Vaidya and Gibbon (1991).

One of the lessons that can be learned from past 'successes' in FSR is that FSR activities have to be seen in a broad framework of rural science and technology promotion. In cases where new methods and approaches have been developed, local actors have been essential innovators in both the development of relevant methods and the way R&D is organised and managed. When there have been 'problems', these innovators have made a commitment to finding the room to manoeuvre in addressing the issues. In being institutionally innovative they have had to:

a.   take risks;
b.   find resources;
c.   influence policy development in their own institutions and possibly in a wide range of research and development agencies.

There are no clear lines between science on the one hand and technology promotion on the other. Also, within science there are no clear lines between basic, applied and adaptive research. In the case of social science theory, the rural surveys and detailed reports of village life in Bangladesh clearly showed that poor people would save resources if institutional opportunities were available. This empirical evidence threw into question the assertions of many economists that poor people have no propensity to save. In the case of plant breeding, the continuous field-level diagnosis of problems, a basic type of research that is fundamental to plant breeding, was shown to need the skills of the most experienced of breeders.

4.2   Defining the Intended Benefits of Research and Development

In all the cases reviewed researchers were not passive in a political sense, but active in their commitment to applying their research skills to benefiting specific groups of poorer people in rural areas. There was no question of developing methods, techniques and institutions that were in some sense 'neutral'.

4.3   Action Research and Rural Poverty

There is a long history of action research which recognises that science is never independent of the society in which it takes place[15]. Two main complementary and inter-related streams can be seen in action research:

a.   <u>Monitoring institutional and technical innovations</u>. These are surveys to monitor, learn from and screen innovations from rural areas.

---

[15]   For a discussion of action research see Castillo (1972); Whyte and Boynton (1983) and Wood and Palmer-Jones (1990).

b. **Rural development experiments.** This is the systematic planning and implementation of rural development experiments.

The rapid rural surveys in the early 1970s in Bangladesh are examples of researchers continuously monitoring and learning from a whole range of innovators in rural areas. The development of the Grameen Bank represents a type of rural development experiment. This is an example of an 'experiment' which did not take place in a 'social laboratory', but in the reality of the existing political and institutional environment.

One of the reasons that FSR has, in some situations, not been able to address the identified technical and institutional problems of rural people, is that the approach has been restrictive - concentrating on the problems of individual farmers in representative groups rather than on issues of agrarian change. A reluctance to address these topics has been a major impasse for many in the FSR fraternity. At a simple level, even the management of cropping patterns frequently brings up property ownership and economic externality issues. For example in Ranchi, India, in the mid 1980s, discussions with poorer farmers led to the conclusion that some winter fallow land which was used for grazing cattle might be better used under a winter crop of pigeon pea. However, if some farmers planted pigeon pea it might be grazed by somebody else's cattle unless means were found to protect the crop. Immediately this led to questions over the ownership of the land and the cattle, and who would benefit and lose from a new method of local resource management. At the time, the FSR group felt this was outside their FSR brief. It was seen as a 'development' issue.

It is important to note that while some FSR practitioners have been reluctant to address issues of common property management etc, there have been researchers who have worked in this area. Interestingly, it appears to be researchers in NGOs who have made substantial progress in this work. For example the landless labourer irrigation-pump group research in Bangladesh is a well documented piece of action research (Wood and Palmer-Jones, 1990). This study clearly documents the political/economic nature of institutional research and the way it interacts with the political context. A lesson from India as regards action research is reflected by a review of many operational research projects in 1986 (Desai and Patel). One of the major findings was that the _research_ projects were frequently seen by many of the actors involved as _extension_ and _development_ projects rather than as _research_ projects. This attitude was reinforced by social science research policy and natural science research policy which frequently undervalues this type of academic research. For those interested in strengthening action research it would appear that attention needs to be focused on ways to change policy in the social and natural sciences.

4.4 Multiple sources of institutional innovations and institutional transfers

It appears that new methods, and new institutional arrangements (whether formal or informal), always evolve in the historical context of the local political and institutional environment. This political economy view of institutional change runs quite contrary to views that widely adaptable 'packages' or '"blue prints' of new methods and institutions can be developed 'outside' and transferred 'into' a new setting. When institutions from 'outside' are transferred into a local situation, at best it can be said that local actors have assessed the relevance of the exotic institutional form, unpacked the components and used those bits which are useful. At worst the apparent transfer of new methods and institutions might represent the undermining of useful local institutions. An apparent transfer from 'a centre of research' might also be part of a process of giving an external name to an institution that already existed locally and had been observed.

This lesson is critical for policy makers, politicians, and planners as it recognises that in science and technology various actors are competing for funds, prestige, etc, and the need to label products and claim originality is part of the normal and imperfect practice of science. This is one of the reasons why the monitoring and evaluation of research is so difficult[16].

5. SUGGESTIONS FOR STRENGTHENING INSTITUTIONS TO REDUCE RURAL POVERTY

5.1 Recognition of the Political Nature of FSR in Rural Development:

The first lesson from experiences of FSR practice is that FSR is not in any sense politically 'neutral.' There will always be groups who will gain more than others, and generally some groups who will lose as a result of different FSR strategies. Where FSR has been effective, the practitioners, have found room to manoeuvre by mobilising resources and being institutionally innovative in the political setting in which they work.

An implication of this is that the relationships between institutional actors, economic policy and FSR practitioners will have to be a central part of FSR analysis in the future. A practical implication of the political economy approach is that 'outside' donors and 'experts' who want to help and strengthen long-term local research and development capabilities to help the rural poor, will have to give primary attention to learning from local people who are already working on institutional innovations to apply science and technology to reduce rural poverty.

---

[16] For a discussion of these issues see Biggs and Farrington (1991) and Biggs (1990).

5.2 Increasing the Range of FSR Analysis to Include the Ownership and Management of Common Property by the Rural Poor

In the past, most mainstream FSR has concentrated on small-farm issues. This has resulted, in some cases, in the neglect of externality problems and neglect of the problems of rural people with few resources and no power. In the future these issues will have to be addressed directly by FSR analysts. As there are many 'action research' practitioners who have been working on these common property topics for many years, it is suggested that FSR practitioners turn to these experiences as they become involved in very complex institutional analysis. One of the implications of this broadening of the interests of FSR work is that more political scientists, macro-economists and rural sociologists will need to be involved.

5.3 Caution in the Use of 'Ideal' Models and Manuals for FSR

The field of FSR is full of flow-charts and organograms, diagrams and manuals of the way FSR should be organized and managed. While these may be useful to practitioners in thinking through possible courses of action, the evidence of successful FSR in the past is that practitioners, whether research-minded, innovative farmers, policy makers, or social or natural scientists; have generally created their own methods, manuals and institutions, in a manner relevant to their needs and environment. The conclusion to be drawn from this is that those concerned with promoting and strengthening FSR with a poverty bias, will have to give more attention to encouraging practitioners to develop, strategies and manuals that are appropriate to and innovative in their own circumstances. And in the process using language and concepts that are locally relevant and suitable for the actors involved.

5.4 Increased Use of Political Economy and Institutional Analysis Methods and Techniques

A fourth implication is that there needs to be more use of political economy approaches, techniques and methods in FSR. This involves the use of such techniques as determinants diagrams, linkage and functional analysis, interest group pay-off matrices, and political economy chronology tables. When determinants diagrams are used in ex-post analysis they are useful in helping to identify which interest groups and actors influenced the direction of institutional changes in the past. When used in a planning mode they can help those who are drawing up plans to identify actors who might gain or lose from specific changes. The diagrams help focus attention on the specific political and institutional issues that have to be addressed in a given planning situation; for example, which interest groups will be against specific policy changes, and what measures are proposed to address these issues. These methods could help reduce the chances that infeasible policies and plans are drawn up[17].

5.5 Broadening the View of Democratic Participation.

Considerable attention has been given in FSR documents to issues of promoting farmer participation. This has proved a very difficult objective to implement. However, it is not only farmer, and landless labourer participation that is difficult to achieve in hierarchical social structures, but it is equally difficult to have junior staff participate in the decision-making bodies of research and extension organisations. Those cases in which FSR approaches have been 'successful' appear to correspond with cases where more democratic egalitarian systems of organisation were in place. These are systems where there was decentralisation of research and extension activities in the allocation of responsibility, power and resources. The flexible use of funds allowed local innovators to create methods and institutions relevant to local needs. An important lesson for FSR policy is that a broad view of democratic participation and the decentralisation of power and resources needs to be encouraged.

5.6 Learning from Practitioners.

The examples in this paper are just a few of the multitude of institutional innovations which might be relevant to those involved in the strengthening of institutions to reduce poverty. Possibly one of the most important lessons is that practitioners might learn from one another, and that information networks between practitioners can encourage this. These might take the form of newsletters, visits to each others' locations, practitioner-workshops, or practitioners reflecting and writing about their experiences and draw out lessons for others. While much can be gained from the work of those who synthesise experiences from across a wide range of situations, there is always the problem that a great deal of institutional and political detail has to be left out of such analysis. The implication of this is that priority should be given to putting practitioners into direct contact with one another.

---

[17] For an example of the use of a determinants diagram in ex-post analysis see Biggs and Griffith (1987). Linkage analysis was used in the ISNAR study of client-orientated on-farm research (Merrill-Sands, et al. 1986). For examples of the use of interest group pay-off matrices see authors in Stewart (1986) and Biggs (1978). Clay and Schaffer (1984) describe a broad framework for the analysis of agricultural and rural development policy.

REFERENCES:

Agarwal, B.D. and others, 1978, 'Maize On-Farm Research Project 1978 Report', Govind Ballabh Pant University of Agriculture and Technology, Pantnagar, December.

Ahmed, N.U., 1975, Field Report on irrigation by Handpump Tubewells, (cyclostyled report). Dhaka: USAID.

Appa Rao, A., 1986, Evaluation in the Indian Agricultural Universities, pp. 46-50 in Daniels, D. (ed.), Evaluation in National Agricultural Research: Proceedings of a Workshop held in Singapore, 7-9 July 1986, IDRC, Ottawa, Canada.

Ashford, T., and S. D. Biggs, 1991, Dynamics of Rural and Agricultural Mechanisation: the role of different actors in technical and institutional change, Paper prepared for the Development Studies Association Annual Conference, Swansea, 11-13th September, 1991, 25p

BARC, 1975, Proceedings of the Workshop on Appropriate Agricultural Technology, Bangladesh Agricultural Research Council, Dhaka.

Biggs, S.D., 1978, Planning rural technologies in the context of social structures and reward systems, Journal of Agricultural Economics, Vol.XXIX, No.3, pp.257-274

Biggs, S.D., 1983, Monitoring and control in agricultural research systems: maize in Northern India, Research Policy, Vol.12, pp.37-59.

Biggs, S. D., 1990, "A Multiple Source of Innovation Model of Agricultural Research and Technology Promotion", World Development, Vol.18 (11), pp. 1481-1499

Biggs, S. D. and J. Farrington, 1991, Agricultural Research and the Rural Poor: A Review of Social Science Analysis, Ottawa: International Development Research Centre

Biggs, S.D., Edwards, C., and Griffith, J., 1978, Irrigation in Bangladesh: On Underutilised Potential, Discussion Paper No.22, School of Development Studies, University of East Anglia.

Biggs, S.D., and J. Griffith, 1987, Irrigation in Bangladesh in Frances Stewart (ed.) Macro-Policies for Appropriate Technology in Developing Countries, Colorado: Westview Press, pp.74-94.

Castillo, G., 1972, Research and the Action Program, A/D/C, New York, 16 pp.

Chambers, R., and Jiggins, J., 1987, "Agricultural Research for Resource Poor Farmers Part 1: Transfer-of-Technology and Farming Systems Research," Agricultural Administration and Extension, Vol. 27, (1), pp 35-52.

Chambers, R., and Jiggins, J., 1987a, "Agricultural Research for Resource Poor Farmers Part II: "A Parsimonious Paradigm," Agricultural Administration and Extension, Vol. 27, (2), pp 109-127.

Clay, E.J. and Schaffer, B., (eds.), 1984, Room for Manoeuvre: An Exploration of Public Policy in Agricultural and Rural Development, Heinemann.

Desai, D.K. and N.T. Patel, 1986, Agricultural Research Management, New Delhi, Oxford and LBH Publishing Co.

Dixon, J. M., 1990, Ways Forward for the Farming Systems Approach in Asia, Paper prepared for the Farming Systems Research and Extension Symposium, Asian Institute of Technology, Bangkok, Thailand

FAO, 1991, Expert Consultation on the Institutionalization of Farming Systems Development - Background Note, Rome: FAO, 6p

Friedrich, K. and M. Hall, 1990, Developing Sustainable Farm-Household Systems: The FAO Response to a Challenge. in Farm Household Analysis, Planning and Development, Proceedings of a Caribbean Workshop, J. Seepersal, C. Pemberton and G, Young (eds), The University of the West Indies, Nov. 1990

Gupta, A.K., (Coordinator), 1988, Draft Recommendations National Workshop on Management of Research on Rainfed Regions December 13-15, 1988 Hyderabad, Centre for Management in Agriculture, Indian Institute of Management, Ahmedabad. 17p

Horton, D., 1984, Social Scientists as Agricultural Researchers: Lessons from the Mantaro Valley Project, IDRC, Ottawa, Canada.

Horton, Doug and Gordon Prain, 1989, "Beyond FSR: new challenges for social scientists in agricultural R&D," Quarterly Journal of International Agriculture, Vol. 28 (3/4), July-Dec. pp. 301-314.

Hossain, M., 1988, Credit for Alleviation of Rural Poverty: The Grameen Bank in Bangladesh, Research Report 65, International Food Policy Research Institute, Washington, D.C.

Kayastha B.N. and Mathema S.B. and P. Rood, 1989, Nepal: Organisation and Management of on-farm Research in the National Agricultural Research System,

OFCOR Case Study No. 4, The Hague: International Service for National Agricultural Research (ISNAR). 159p.

Maurya, D.M., 1986, On-farm rice research for resource poor farmers of Eastern Uttar Pradesh, Report 1984-86, Dept. of Genetics and Plant Breeding, Narendra Deva University of Agriculture and Technology, Narendranagar, Faizabad.

Maurya, D.M., Bottrall, A., Farrington, J. 1988. "Improved livelihoods, genetic diversity and farmer participation: a strategy for rice breeding in rainfed areas of India". Experimental Agriculture, Vol. 24, pp311-320.

Merrill-Sands, D., 1986, "Farming Systems Research: Clarification of terms and concepts", Experimental Agriculture, Vol. 22, pp.87-104.

Merrill-Sands, D., S. D. Biggs, S. Kean, S. Poats, J. McAllister, E. Moscardi, and S. Ruando, 1986, Methodology for the ISNAR Study on the Organisation and Management of On-Farm Research, The International Service for National Agricultural Research, The Hague, Netherlands. 69p.

Merrill-Sands, D., P. Ewell, S. Biggs and J. McAllister, 1989, Issues in Institutionalizing On-farm Client-oriented Research: A Review of Experiences from Nine National Agricultural Research Systems, Quarterly Journal of International Agriculture Sept., Vol. 3 (4), pp. 279-300.

Merrill-Sands, D., P. Ewell, S. Biggs, R.J. Bingen, J. McAllister, and S. Poats, 1989a, Institutionalized on-farm Client Orientated Research: Reflections on the Experiences of Nine National Agricultural Research Systems. The Hague: International Service for National Agricultural Research.

Orr, Alastair, Nazrul Islam, ASM, and Barnes, G, 1991, The Treadle Pump: Manual Irrigation for Small Farmers in Bangladesh. Dhaka: Rangpur Dinajpur Rural Service.

Stewart, F., (ed), 1987, Macro Policies for Appropriate Technology in Developing Countries, Westview Press, Boulder, Colorado,

Vaidya, A. K. and D. Gibbon, 1991, Survival and Sustainability in the Mid-Western Hills of Nepal, Paper presented at the 11th Annual Farming Systems Research-Extension Meeting organised by the Michigan State University, East Lancing, Michigan, USA, October 5-10, 1991. 19p

Vishnumurthy, T., 1987, Feed-back from Operational Research Projects and Reorientation of Research Programmes, Hyderabad: Central Research Institute for Dryland Agriculture, 12p.

Whyte, W.F. and D.Boynton, (eds), 1983, Higher-Yielding Human Systems for Agriculture, Cornell University Press, Ithaca.

Wood, G. (ed), (1985), Labelling in Development Policy, London: Sage.

Wood, G., and Richard Palmer-Jones, 1990, The Water Sellers: A Collective Venture by the Rural Poor, London: IT Publications

Yunus, M., 1977, Story of a Deep Tubewell with a Difference: A Report on Osmania School Purba Beel Tubewell, Raozan, Chittagong, Locally Sponsored Development Programme Series: Report No. 28, Rural Studies Project, Department of Economics, Chittagong University, Bangladesh. (Published by the Ministry of Cooperatives, Local Government and Rural Development, Government of the People's Republic of Bangladesh), 84p.

III

ADOPTING FARMING SYSTEMS APPROACHES IN EDUCATIONAL INSTITUTIONS

Overview and Synthesis

Most experience with farming systems approaches in the agricultural sector of developing countries has been in association with donor agencies, which usually instigate and finance such efforts. Apart from raising the question of sustainability once donor support comes to an end, these experiences are generally set in a project context. They do not, therefore, fully respond to the need to embed the systems perspective into the work and procedures of national development institutions. One powerful contribution towards achieving this aim is discussed in the present chapter; it concerns the development of manpower with appropriate knowledge and skills.

Some universities in the Third World have addressed this challenge by seeking to introduce farming systems aspects into their teaching and research work. Despite the obvious advantages of initiating farming systems training prior to starting upon a professional career, examples are few and of recent origin. Success in this area is of the greatest importance, however, since re-orientation seminars and other types of in-service training are not a satisfactory substitute for a basic professional education in which the systems perspective is integral to the overall syllabus.

Where teaching reforms have been attempted, the underlying rationale has been to expose students to both the underlying concepts and basic techniques of farming systems in order to:

- introduce a holistic perspective to replace the reductionist, partial approaches followed by most agricultural courses at university level,

- use and develop the experience of students to create an understanding of the main elements and interactions of a given farming system,

- link development objectives, constraints and solutions at the micro-level to institutional and policy issues,

- understand system changes and adjustments within a longer-term context, and

- quantify relationships, problems and solutions in order to estimate the impact of change.

Some instances of confusion have been reported with respect to variations in terminology and choice of techniques. Nonetheless, it is widely agreed that the teaching of farming systems should emphasize a common core of essential components. Prominent among these is the importance of technical knowledge applied within the multi-sectoral context of rural development. The holistic nature of farming systems analysis means that an inter-disciplinary approach is basic to the curriculum. Another basic element is the importance of the micro-macro continuum and the interaction of a given system, at any point of the continuum, with its environment. Above all, the teaching seeks to develop a problem-solving approach in place of pure academic study. This involves the development of a close relationship between students, farm families and government and non-government institutions.

The attempt to extend the educational environment to the field situation has given rise to several problems. The most obvious of these arise from the extra burden that such activities impose on limited budgets and transport facilities.
A related problem is the development of a stable, productive relationship between all parties concerned, so that field experience can be established as a regular feature of the curriculum. This problem is partly the result of the present weakness of links between universities, agricultural development institutions and surrounding rural communities.

The second major group of problems arises from the scarcity of suitably trained and experienced manpower needed to teach a farming systems perspective. Few specialists are attracted to become involved in farming systems teaching and research. Since it does not attract much academic prestige, incentives to change are minimal. Even those teachers who are motivated to change find it difficult to acquire the necessary training while sustaining a full teaching load. In addition, training materials are extremely limited and venues for publishing farming systems research fairly restricted.

This chapter focuses on a range of experiences in introducing farming systems concepts in a university setting. The experience of the University of Agriculture at Morogoro in Tanzania concentrates upon improvements in undergraduate teaching. A two-phased approach, similar to that advocated in the paper by Doppler and Maurer, was initiated in 1988. Activities centred upon teaching the systems concept in selected courses, while making suitable reference material available. Sensitization seminars, study visits and training workshops were also organized with the aim of preparing staff for a new curriculum to be introduced in the second phase.

The experience of Brawijaya University at Malang in Indonesia, concerns an attempt to introduce an agro-ecosystems approach to the research programme. Although the subject was introduced at a workshop in 1984, a full programme of field research started only in 1989, in collaboration with the University of Wageningen. The initiative is still in project form and only 7 members of staff are engaged in the research, but several important lessons have already been learned.

The Agricultural University (ViSCA) in the Philippines has promoted research and development with a farming systems perspective since 1981, in collaboration with the Regional Department of Agriculture. A separate unit was formed in 1987, by creating a multidisciplinary team to draw upon the resources of other colleagues in the University in order to help resource-poor farmers. This was done in full recognition of the fact that separate units can become divorced from the mainstream of the work of the parent institution. It is this recognition that makes the paper so interesting.

FARMING SYSTEMS DEVELOPMENT IN ACADEMIC TRAINING

by

Doppler W.[1] and M. Maurer[2]

1. THE NEED FOR UNIVERSITIES TO CONTRIBUTE HEAVILY TO FSD

The objective of farming systems development (FSD) is the sustainable development of farm household systems. Decisions at farm and household level have significant impacts on national and regional development. Conversely, decisions at national and regional levels (in both the public and private sectors) have important effects on farm household systems. FSD deals with the decisions taken at farm and household level, but includes the consequences of macro level decisions and hence has a micro as well as a sector component. From the standpoint of rural development, FSD is multi-sectoral.

The FSD approach is based on the fact that farm households make decisions concerning their particular members and holdings and with respect to external relationships according to their personal needs and objectives. The intention is to come as close as possible to the reality of decision-making. FSD takes account of: the complexity of sectoral and micro level objectives, which are often conflicting; an understanding of the farm and its household as instruments; the dynamics of processes; and lastly relationships with the external environment. The above holistic situation calls for a multidisciplinary concept.

The FSD approach is not yet being applied in teaching and research in university agricultural faculties in developing countries. Individual academics may see a need for such a concept, but the vast majority still practice the traditional approach. This frequently overemphasizes: the short term, a very small number of disciplines, and a single crop or type of animal. The traditional approach is often orientated to a particular agricultural commodity. Finally, the medium and long term objectives of human beings are often neglected or completely ignored by the traditional approach.

---

[1] Professor for Farming Systems Economics in the Tropics, Head of Department, University of Hohenheim, Germany

[2] Agricultural Economist in an FSD Research Team at the University of Hohenheim, Germany

If FSD is adopted as a key strategy for development of agriculture by the governments, farmers and public and private sector infrastructural agencies of the Third World, there will be an urgent need for universities to respond positively. In this event FSD will become a vital area of activity in faculties of agriculture and related subjects. Extension services, non-governmental organizations (NGOs), ministries and other institutions recruit their professional staff mainly from universities, which, almost alone, will have to act as the springboard for implementing the FSD concept. The training of extension personnel is both a short and medium term activity, but it will have no sustainable impact if the universities do not replace the traditional approach with that of FSD. Accordingly, this paper focuses upon the need for, and determined acceptance by universities, of FSD. The fact that in terms of staffing and other resources teaching plays a larger role than research will be kept well in mind.

2. OBJECTIVES OF FSD TRAINING

The five main aims of FSD academic training are as follows:

a) **To familiarize students with the philosophy of the FSD approach**
Creating awareness and familiarizing students with the FSD approach are prerequisites for its incorporation into university curricula. Many students have grown up in the countryside and are conversant with the nature and complexity of decisions which have to be taken on farms and elsewhere in rural areas. These students bring a wide range of vital knowledge and experience of practical day-to-day farm and other rural problems. This pool of knowledge and experience is hardly tapped in most universities in the Third World, since instruction continues to focus on the traditional, narrow, discipline-orientated, short-term approach.

b) **To train students in the methodology and technique of FSD**
This training is needed to familiarize students with the principles of the methodology, as well as to appreciate the potential gains which can be expected to follow from the adoption and application of FSD. This instruction will provide the basic knowledge for the application of FSD methodology to extension work and research, as well as an instrument for policy analysis.

c) **To train students in the application of the FSD approach to research**
Training in FSD research is needed

  (i)   to provide personnel for research,
  (ii) to be able to design and propose national FSD research programmes
  (iii) to persuade decision-making bodies to provide the resources required for carrying out such research programmes.

d) **To train students in the application of the methodology of the FSD approach to extension work.**
   This training is needed

   (i) to strengthen the extension services by providing FSD-trained personnel,
   (ii) to allow extension personnel to test and define farmer-orientated solutions to practical problems, thereby leading to increased efficiency, particularly in terms of higher adoption rates, and
   (iii) to expedite the improvement of the living standards of farm families.

e) **To train students how to use the FSD approach as an instrument for policy analysis at both national and regional levels**
   This training is needed

   (i) to supply national and regional policy-making and implementing bodies with FSD-trained personnel and thus strengthen the holistic understanding and FSD approach of policy-makers and administrators,
   (ii) to develop administrative concepts and political programmes and solutions based on less-incomplete, but more reliable information, leading to improved decision-making at the micro level; this will take full account of the likely reaction of farmers to institutional and political decisions, and
   (iii) to enable policy-makers and administrators to foresee the likely direct and indirect results of their actions as accurately and early as possible.

Since in many third world universities fsd has not been introduced and in some is not even known, short-term objectives should focus on teaching staff as follows:

a) To create awareness by the teaching staff of the relevance and need for FSD.
b) To familiarize teaching staff with the philosophy, methodology and application of FSD under practical conditions.

3. BENEFITS EXPECTED

In the course of time, the more university training takes place the less training of extension and ministry staff will be required when the graduates come to take their professional positions. It is for this long-term aspect that farming systems training at the universities should be given a high priority and regarded as a long-term investment.

Benefits expected to follow from the introduction of the FSD approach to university training are as follows:

a) Strengthening research, extension, and teaching institutions through a common philosophy and strategy, thereby leading to better use of resources.
b) Increasing the efficiency of development projects by testing, evaluating and providing solutions which are more readily accepted by farmers. Increasing farmers' motivation through participation.
c) Improving the situation of farm families by orientating activities, measures and proposals towards their needs, objectives and problems.
d) Providing better information (e.g. by making it more comprehensive, realistic and reliable) for policy-makers and public and private sector investors.
e) Ensuring that research programmes are, generally speaking, holistic, rather than isolated, narrow and short term.

The overall benefits wil. include a long-term improvement in the living standards of farm and other rural families as well as better resource use by the agricultural sector.

4. RATIONALE AND METHODOLOGY OF AN FSD CURRICULUM

An FSD curriculum in an agricultural faculty is meant to provide the logical structure of students' teaching and training. The rationale of an FSD curriculum should be:

a) To use and develop the experience of students to understand farming as a system which aims to help solving problems of human beings. The family is the core and the farm and the household are the instruments for achieving the objectives of the family.
b) To understand both the complexity of farmers' real world and of decision-making under long-term perspectives. This includes the need for, and sustainability of resources. Since a single discipline cannot cover what is needed, interdisciplinary work is required.
c) To define research problems and develop and test solutions by integrating the families whose problems are to be solved. Participation should play a central role.
d) To link objectives and solutions at micro level to the constraints and to institutional and policy issues. Any development at micro level will have to be understood in the context of social development.
e) To quantify relations and problems so as to show and justify the impact of what is being proposed.

Based on the above overall concept of a farming systems approach a university undergraduate curriculum should be structured along the lines of Figure 1 and conform to the following conditions:

- the interdisciplinary systems approach will be employed throughout the studies;
- in addition to the systems approach a minimum of basic technical knowledge of the relevant disciplines will be required and this will be applied by means of the FSD approach;
- the systems approach will be dealt with from both the sectoral and micro standpoints;
- the sectoral systems consist of resources, the environment, institutions and behaviour;
- the farm/household system will be dealt with from a research point of view (to develop and test solutions) and from an extension point of view (to transfer of solutions to farms and households).

Fig. 1 An overall concept for an FSD curriculum at universities

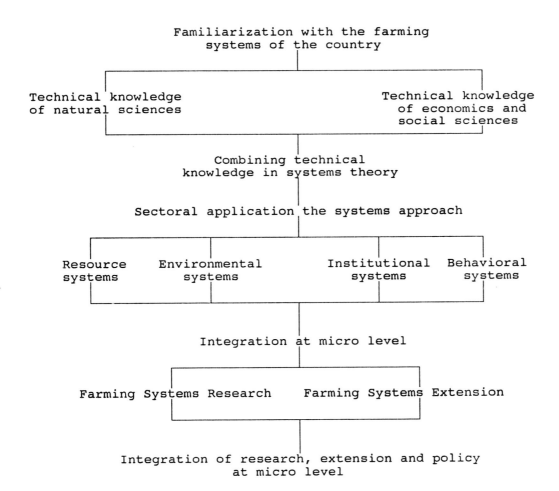

The methodology of FSD to be dealt with in the undergraduate curricula should cover all relevant areas. Any specialization in the methodology should be avoided, since during study time it is not known whether a student will, after qualifying, work in research, extension or administration. The methodology should cover all facets of FSD, including Farming Systems Analysis (FSA), and Farming Systems Planning, Monitoring and Evaluation (FSP). The purpose of FSA is to understand and explain past development and the present situation as a basis for designing possible solutions to problems in the future. The object of FSP is the development and evaluation of long-term solutions within the frame work of a Systems Impact Analysis (SIA). The purpose of SIA is to determine the impact of new solutions on the objectives of farm families ("with" and "without" principle). Farming Systems Research (FSR) is often considered as the part of FSD which will provide the inventions and the innovations which will be disseminated by Farming Systems Extension (FSE) institutions. Quantitative as well as qualitative methods should play a central role but quantitative methods will be given preference whenever possible. In areas where quantification is not possible, the qualitative approach is the most relevant methodology of information-gathering.

The introduction of FSD in a curriculum will have the following implications for teaching methods:

a) Interdisciplinary teaching is needed. This requires intensive collaboration between teachers. Obviously, the traditional teaching methodology of e.g. three separate courses given by separate teachers from different specialized backgrounds in soil science, plant nutrition and crop production is different to the alternative of an interdisciplinary course in soil-water-plant-relationships.
b) The study area approach, where students work with farmers and extension personnel in an outreach area of the university calls for extremely communicative teaching methods. Learning by discussion and communication plays a central role.
c) The integration of sample farmers and extension personnel into the traditional student-teacher-relationship is necessary.

5. IMPLEMENTATION OF AN FSD CURRICULUM

A pre-condition for the successful implementation of an FSD curriculum is the collaboration of students, farm families and the extension institutions. This is to relate the methodology offered in undergraduate courses to the farmers' real world and to those who intend to find solutions to farm families' problems. This should be assured by the following:

a) Students will undergo periods of carefully designed and monitored practical training before or during their course of instruction. The training will be given by agencies such as farms, extension services, banks, co-operatives or in marketing, processing and other businesses. The objective will be to familiarize students with day-to-day problems in rural areas, to understand differences in farms and households, learn how farmers see their problems and make decisions, and appreciate the complexity of farmers' objectives and decisions.

b) Parallel with the courses taught at the university, students should discuss the topics of the lectures with rural families through excursions, farm interviews and group discussions in villages. The objective will be to ensure a close linkage of the subjects taught to reality.

c) In defined geographical areas, students should carry out individual studies (e.g. seminar papers and master theses). The objective will be to gain experience in how to conduct interviews, organise data collection, approach people, and learn how to apply the methodology under practical conditions and understand the global context of rural problems. Individual studies will be designed so that each forms part of a wide problem area.

d) Study results will be presented in university seminars and in meetings with the farmers concerned. The objective will be to learn how to interpret the results of analysis and planning, gain experience in carrying out scientific discussions and in liaising with farmers.

e) In the study area, students will be trained in extension and communication activities, especially in defining problems, selecting solutions to be offered to farmers and obtaining their reactions. The objective will be to understand the need for farmers' participation and appreciate the impact of extension.

The implementation of FSD in a curriculum faces the problems of limited staff and financial capacity. For those reasons, it is suggested that FSD should be introduced in stages over time. The pace of change should depend on the status of existing FSD courses in a curriculum as illustrated in the following simplified two-stage approach:

a) Integrating farming systems components into the existing curriculum of an agricultural faculty

In this stage only the most important FSD components are integrated. An example is given in Table 1. In a typical curriculum of four years training about 37% of the courses could be offered in the field of FSD. This can be considered as the minimum to ensure that graduates are able to apply the farming systems approach.

The advantages of integrating FSD into the existing curriculum structure are that the staff members can be partly be utilised with little further training and that in the short and medium term it will be less costly, since less training will be required. A disadvantage is that students may be confronted with two opposing approaches which might lead to conflict since some teachers may choose the keep to the commodity-orientated approach, which is often easier to understand.

b) Establishing an "FSD-Approach" Curriculum

A tentative, detailed curriculum covering a four-year undergraduate course is given in Table 2. The main characteristics are as follows:

i) The courses will, equally weighted, cover all areas relating to farm production, processing, marketing, storage, home consumption, household, family, social development and socio-cultural relations, anthropology and farmers organizations as well as economics, institutions and policy.

ii) Based on the FSD approach, the courses include interdisciplinary concepts and the dynamics of the local farming systems, and they will require the participation of farm families. The FSD approach is reflected in the fact that e.g. soil-water-plant relationships replace soil physics, soil chemistry, plant nutrition and plant physiology.

iii) The courses will focus on problems specific to the country and examples will be provided in selected study areas in order to extend knowledge about the local farming systems.

iv) A large number of individual studies need to be carried out in the university's main catchment area, so as to learn to apply the methodology of farming systems research and development, especially in continuous discussions with farmers and their families. This is of especial importance for the methodology of information-gathering as well as the on-station, on-farm and on-household testing of results.

v) Close collaboration with farmers will familiarize students with the potential gains from, and problems of, integrating farmers in the definition of research problems, the testing and evaluation of solutions as well as in the relevance of the decisions about the dissemination of solutions within an extension service. Experience gained in applying the methodology of research priority-setting will prepare students for future extension activities.

vi) FSD is the methodology which bridges the gap between research and extension (development) and therefore plays a central role in a curriculum from a theoretical as well as an empirical point of view. The application of FSD requires close collaboration with farm families and the extension personnel. Continuous impact analysis in a given area will provide an on-going continuous learning process in and for future generations of students. The implementation of the proposed measures on selected farms over time will provide benefits from the long-term impact of the proposals and expose difficulties which may arise under practical conditions.

vii) Close collaboration between students and extension personnel will provide the undergraduates with practical experience in all the main aspects of agricultural and other rural advisory work.

The advantages of the approach given in Section 5 a) over those set out in Section 5 b) are that in the long-term the former approach is less costly, the probability of success is greater, and it is more likely to help raising the standard of living of farm families and contribute to the success of development activities.

This is partly because the activities will be evaluated according to the degree of acceptance by farm families before they are disseminated through the extension service.

The disadvantages of the approach set out in Section 5 b) are that it requires an increased administrative input to organize the collaboration between students, farmers and extension staff as well as additional transportation. Finally, the Section 5 b) approach requires, in many cases, a large amount of training of university staff.

6. CONSTRAINTS AND REASONS FOR FAILURE

The most relevant constraints to the introduction of FSD in universities can be summarized as follows:

a) **Curricula**
The inflexibility of existing discipline-oriented curricula. In many countries a curriculum may only be changed if a majority of the academic staff, at every level, is in favour. In some cases the approval of the relevant ministry is also needed.

b) **Teaching staff**
- The lack of awareness of the relevance of FSD on the part of many academics. In many cases the majority of the university staff have little or no knowledge about the FSD approach. Often the cropping systems, on-station or on-farm testing are regarded, erroneously, as the FSD approach. Academics who have been trained in the traditional way, simply do not know, let alone, appreciate FSD approach. Concern that new concepts have to be learned, may hinder the acceptance of FSD.
- Training of university staff is required. Given the willingness of the staff, the large amount of time needed to learn and experience the FSD approach might not be available, because of the given teaching load and research commitments. In addition, intensive and comprehensive training courses are costly and require a large input of external personnel.

c) **Financial and managerial resources**
A farmer- and field-related curriculum imposes a large running cost component on the universities. This is partly due to transportation from university to farms and villages as well as for on-farm and on-household tests. (These costs are modest per undergraduate-year and are believed to be more than offset by the improved quality of graduates.)

d) **Willingness of farmers and extension personnel**
- Willingness to collaborate with students and university staff. Finding study areas where farmers are willing to collaborate may prove difficult in the

initial stages. In the case of on-farm and on-household research and tests one has to find typical farms and families and make it clear that even when the solutions are put into practice they may not always be successful. The risk of failure will not be burdened on farmers' shoulder.

- Readiness of extension services to collaborate with students and university staff. A minimal acceptance of the philosophy of FSD is a precondition for collaboration of the extension services. In addition, it must be clear that this approach and collaboration are designed to support, and in no way to undermine the extension service.

e) **Teaching material**
There is a lack of teaching material in FSD in the language used in the courses. The tendency of many countries to introduce a local/national language has led to an abrupt curtailment of supplies of international textbooks and other teaching materials. This is true even for FSD textbooks in widely used languages like Arabic.

f) **Capacity to train teachers**
To train teachers in universities in developing countries in FSD will require a large number of experienced teaching staff. If only ten universities sought to train inexperienced staff at the same time, the required capacity would not be available.

7. SUMMARY AND CONCLUSIONS

Training at universities is a long-term investment with wide-ranging effects on society. Extension services, research institutions, NGOs, ministries and other institutions recruit their staff mainly from universities. The introduction of FSD to universities consequently provides the basis for a country-wide introduction of this concept.

The objectives are to familiarize students with the philosophy of the FSD approach, to train students in the methodology and techniques, to train students to apply the methodology to research and extension, and the use of the approach as an instrument of policy analysis at regional and nation levels.

The benefits expected from the adoption of FSD by Third World universities include: the strengthening of institutions, an increase in efficiency of development projects, the improvement of the situation of farm families and the provision of better information for policy decisions.

The rationale of a FSD curriculum is based on multidisciplinary and multisectoral components and includes the participation of farm families and extension staff in a study area. Consequently, teaching methods have to be adjusted to this interdisciplinary, farmer- and field-oriented and dynamic approach.

The implementation of FSD in universities requires in many cases a sequence of steps which depend on the level of awareness and familiarity of the teaching staff with FSD. For a two-stage process, examples of a curriculum are presented. The integration of elements of FSD into a typical curriculum of an agricultural faculty is considered to be a transition phase which could in a second stage, lead to a complete FSD curriculum. The latter would require a heavy component of university staff training before it could be put into practice.

The most relevant constraints to be expected would be the inflexibility of existing discipline-oriented curricula, the lack of awareness of the relevance of FSD for university staff and the lack of FSD training of university staff. These constraints could, under specific circumstances, probably be surmounted by the provision of external personnel and funds. Readiness of farmers and extension staff in a study area to collaborate and participate in the programme is important, but they ought not to prove serious obstacles. The relatively high running cost components may create budget problems in a university.

The general conclusion of this analysis of the potential contribution of the introduction and institutionalization of FSD in Third World universities can be summarized as follows:

a) In spite of the theoretically ideal situation of a university starting to introduce FSD from the beginning of the training and at an institution with a wide snowball effect, local universities cannot, in many cases, take the lead in introducing FSD in a developing country. The constraints and problems are so important and the possibilities for solutions are practically so limited that reality allows only small steps. In many cases, the Third World universities will follow the introduction of FSD in research and in some cases extension.

b) The introduction and institutionalization of FSD into the curriculum will, in most cases, be slow and, in stages over time. The capacity to effect changes in universities is often much more limited than in research and extension. Awareness workshops, published material, short training courses for teaching staff, the introduction of a few FSD courses, intensive training programmes for teaching staff and more comprehensive training would all be the result of a long-term strategy of institutionalizing FSD in a Third World university.

REFERENCES

(1)   FAO, (1990): Farming Systems Development. Guidelines for the conduct of a training course in Farming Systems Development. Rome.

(2)   DOPPLER, W. (1991): Landwirtschaftliche Betriebslehre in den Tropen and Subtropen. Manuskript. Stuttgart. (Farming Systems Economics in the Tropics and Subtropics).

(3)   FRIEDRICH, K. and HALL, M. (1990): Developing Sustainable Farm-Household Systems: The FAO Response to Challenge. In: Farm Household Analyses, Planning and Development. Proceedings of a Caribbean Workshop. SEEPERSAL, J.; PEMPERTON, C. and YOUNG, G. (eds.). The University of The West Indies.

(4)   WEINSCHENCK, G., 1989: From Subsistence Households to sustainable farming environment systems. In: Quarterly Journal of International Agriculture, Vol. 28, N. 3/4, July - December 1989, DLG, Frankfurt.

In addition: unpublished personal and faculty documents of the authors based on research conducted at the Agricultural University of Malaysia, the University of Nairobi and Egerton, Kenya, and Morogoro, Tanzania.

Acknowledgment

We would like to thank Eric Gibbons for his valuable ideas and comments based on his experiences in universities in Africa and Asia.

Table 1 The integration of FSD components into a typical curricula of an agricultural faculty (courses of 18 hours/week, 2 terms/year)

		credit
YEAR 1		
Physics		2
Chemistry		2
Agricultural chemistry		2
Botany		2
Agricultural botany		2
Agricultural microbiology		2
Genetics		2
Mathematics		2
Statistics		2
Econometrics		2
Systems theory	(FSD)	2
Behavioral theory	(FSD)	2
Basic economics		2
Principles of social development		2
Human civilization and development		2
Principles of community		2
Farming systems in the country	(FSD)	2
Agricultural development in the country	(FSD)	2
YEAR 2:		
Climatology		2
Soil physics and conservation		2
Soil chemistry and microbiology		2
Crop production and crop husbandry		2
Plant breeding		2
Crop physiology		2
Crop protection		2
Agricultural engineering		2
Farm mechanization		2
Processing of food, beverages and others		2
Principles of farming systems research	(FSD)	2
Principles of farming systems development	(FSD)	2
Animal production and husbandry		2
Animal nutrition		2
Animal health		2
Resource systems	(FSD)	2
Environmental systems	(FSD)	2
Institutional systems	(FSD)	2
YEAR 3:		
Principles of economics		2
Micro-economics		2
Macro-economics		2
Farm systems analyses	(FSD)	2
Farming systems planning	(FSD)	2
Household systems analyses	(FSD)	2
Household systems planning	(FSD)	2
On-station and on-farm research	(FSD)	2
On-household research	(FSD)	2
Computers		2
Information systems and processing		2
Project planning and evaluation		2
Marketing		2
Rural development		2
Credit systems and services		2
Agricultural policy		2
Development policy		2
International development		2

Table 1 (contd)

YEAR 4:		credit
Agricultural resources		2
Resource evaluation		2
Food and nutrition	(FSD)	2
Household supply	(FSD)	2
Family needs	(FSD)	2
Family planning	(FSD)	2
Extension and extension services		2
Agricultural extension		2
Communication systems		2
Monitoring and evaluation systems		2
Survey	(FSD)	2
Advanced farming systems development	(FSD)	2
Research priority setting	(FSD)	2
Definition of extension packages	(FSD)	2
Organization of extension	(FSD)	2
Policy papers	(FSD)	2
Research contribution to policy	(FSD)	2
Extension contribution to policy decisions	(FSD)	2

Table 2: A proposal for an FSD curricula courses (18 hours/week, 2 terms/year)

YEAR 1:	credit
Physics	2
Chemistry	2
Botany	2
Zoology	2
Soil-water-plant relations	4
Cropping systems	4
Nutrition-health-animal relations	4
Statistics	2
Systems theory	2
Behavioral theory	2
Human health and nutrition	2
Principles of social development	2
Human civilization and development	2
Principles of community	2
Farming systems in the country	2
YEAR 2:	
Agricultural development in the country	2
Crop rotation and resource conservation	2
On-station cropping systems improvement	4
On-farm cropping systems improvement	4
Mixed farming systems	2
Livestock integration	2
Farm mechanization and investment	2
Processing and marketing	2
Farm and household economics	4
Socio-cultural development	2
Methodology of farming systems approach	2
Farmers' participation	2
Anthropology	2
Introduction to computers	2
Information systems and processing	2

Table 2 (contd)

YEAR 3:	credit
Resource systems	2
Environmental systems	2
Institutional systems	2
Behavioral systems	2
Farming and household systems analyses	4
Farming and household systems planning	4
On-farm and on-household research	2
Project planning and evaluation	2
Marketing	2
Rural sociology	2
Credit systems and services	2
Agricultural policy	2
Computer application	2
Survey in study region	2
Research and development papers	2
Group discussions with farmers	2
Policy papers	2
YEAR 4:	
International development	2
Resource evaluation	2
Food and nutrition	2
Household supply	2
Family needs	2
Family planning	2
Extension and extension services	2
Agricultural extension	2
Communication systems	2
Monitoring and evaluation systems	2
Advanced farming systems development	2
Research priority setting	2
Definition of extension packages	2
Organization of extension	2
Policy papers	2
Research contribution to policy	2
Extension contribution to policy decisions	2
Group discussions with farmers	2

INSTITUTIONALIZATION OF THE FARMING SYSTEMS RESEARCH AND EXTENSION APPROACH IN THE AGRICULTURAL TRAINING SYSTEM IN TANZANIA

by

A.Z. Mattee[1]

1. INTRODUCTION

The Farming Systems approach to research and extension (FSR/E), was introduced in the eastern African region in the mid-seventies, and has undergone a certain degree of evolution in terms of conceptualization, nomenclature, specification of techniques or methodologies involved, and in terms of incorporation (institutionalization) into existing research and extension systems.

However defined or named, it is now widely accepted that the approach is farmer-oriented, problem-focused and incorporates a systems perspective, including a multi-disciplinary approach to the development of recommendations which are compatible with the farming systems, farmers' objectives and preferences.

The current concern with FSR/E has emerged out of the recognized limitations of the conventional research and extension approaches in generating and disseminating appropriate technologies which would have immediate impact on farmer's productivity. This limitation has been mainly because conventionally, researchers and extension workers have tended to focus on specific farm constraints to the exclusion of others (or holding them constant!) which also affect the productivity of the farm. This has in turn resulted in failure to use a managerial or systems perspective to diagnose farm constraints and to develop appropriate technologies. This has also resulted in poor institutional linkages and inadequate policies (Collinson, 1984; Opio-Odongo, 1986; Friedrich, 1988; Norman, 1980).

The approach has emerged as a way of redressing this weakness. It is based on the assumption that:

---

[1] Department of Agricultural Education and Extension, Sokoine University of Agriculture, Morogoro, Tanzania

- Small farmers operate under diverse circumstances
- Small farmers manage systems to meet a multitude of objectives - they often make compromises in order to maximize returns, and to ensure the sustainability of the system (Norman and Collinson, 1985).
- Indigenous knowledge of basic technical information on enterprise production in a given natural environment exists (De Gregory, 1969; Brokensha et al, 1980).
- Small farm systems are dynamic and receptive to change, but tend to be selective and adaptive in their adoption and use of technologies (Merril-Sands, 1986).

The Farming systems approach therefore, enables the researcher to work directly with farmers and extension workers with the objective of:

- understanding better the farmer's circumstances,
- deciding on priority problems of the farmer which need to be tackled in the short term, given the farmer's circumstances and current level of technology,
- developing and testing technologies which will address those priority problems,
- influencing research and extension policies and practices in order to come up with more relevant research and extension programmes.

Friedrich (1988) summarizes it as
> "..... an approach, a state of mind, supported by appropriate methods and procedures....... for informing on and creating an understanding of farm situations, problems and potential interventions at all levels of the development hierarchy, from the policy maker, researcher, manager to the front-line extension agent" (pg. 4).

2. THE FARMING SYSTEMS RESEARCH AND EXTENSION APPROACH IN TANZANIA

The approach was introduced in Tanzania in the seventies by the International Maize and Wheat Improvement Centre (CIMMYT), and later on several other donors including International Development Research Centre (IDRC) of Canada, USAID, FINNIDA etc. funded subsequent initiatives. Of late SIDA and FAO have also began to be involved in Farming Systems initiatives in Tanzania as well as in other countries of the SADCC region. Thus most initiatives have been implemented using the project approach whereby each donor operated in isolation. This has resulted in:

- different terminologies and techniques being developed for virtually the same process,
- lack of continuity in programming and resource commitment,

- Lack of commitment on the part of staff not in FSR/E projects
- fragmentation of effort, particularly the scarce manpower already involved in Farming systems work.

This begs the question, to what extent can the process be firmly accepted and incorporated into the National Agricultural Research and Extension Systems, after donor commitment ends. The rest of the paper discusses an attempt at Sokoine University of Agriculture (SUA) to contribute to the process of institutionalizing the approach in Tanzania.

3. INSTITUTIONALIZATION OF FSR/E TRAINING AT SUA

3.1 Background

Institutionalization of the FSR/E approach simply means the permanent integration of the approach within the National Agricultural Research and Extension systems. Successful institutionalization would presuppose the following:

- a clear demonstration of the usefulness of the FSR/E approach,
- government commitment to integrate the approach into the National Agricultural Research and Extension Systems,
- A clear strategy for integrating the approach into the National systems, including plans of activities, resource commitment, phasing and even structural adjustments.
- a systematic process of developing manpower with the technical skills and attitudes necessary to use a systems perspective in their research and extension work.

It is quite evident that even with government's commitment to institutionalize the approach in the National systems, the most critical factor is the availability of adequate manpower with the necessary knowledge, skills and attitudes.

And yet, so far FSR teams have acquired their skills through various short-term training programmes which have been supported by international organizations, and the building up of FSR capacity into the country has depended almost entirely on foreign assistance. In addition, this training has almost always been tied to specific projects with farming systems components. No systematic training programme existed in the country to ensure that all our future researchers and extensionists will acquire the necessary skills to adopt a systems approach in their work.

The graduates who turn out of Sokoine University of Agriculture are not suited to the organizational structure of the typical farm household that they are supposed to serve. The type of graduates being turned out at B.Sc. level are neither scientists nor agriculturalists. Specialization has tended to narrow the perception of the real farm situation which forms the bulk of small farmer population. Thus, young professionals are only able to use their knowledge in the format they received it at the university; they cannot easily build up new solutions, meeting farmers' demands, by integrating all subjects assimilated during their studies (AAASA, 1975; Huxley, 1976; Moris, 1976; Keregero, 1981; Mattee, 1986, 1988). In order to ensure that the FSR/E approach is institutionalized in the National Agricultural Research and Extension Systems it is necessary therefore for training institutions, those preparing future researchers and extensionists, to adopt a systems perspective in their teaching, so that the students can acquire a holistic view of the farmer situation. Opio-Odongo (1986) identified three benefits to trainees subjected to a systems perspective in their training:

a) an ability to analyze and understand the circumstances and opportunities facing local farmers,
b) acquisition of skills to de. 'e experimental trials from priority problems of local farmers which can improve the likelihood of presenting them with relevant recommendations, and
c) a safeguard against the common tendency of compartmentalizing knowledge and drawing a sharp distinction between research and extension (pg 211).

In addition, a challenge was thrown to the University academics during the inauguration of SUA. The then President of Tanzania Mwalimu Julius K. Nyerere (1984) in his inauguration speech, mentioned that,

> "The major purpose of this University is the development and transmission of skills and practical expertise at the highest level..... This University must be answering to the needs and solving the problems of Tanzanian agriculture and rural life. Its aim must be, first to contribute towards improved production and therefore standards of living for the people who live and work on the land or in connection with the land".

The point is that, the university must endeavour to structure its curriculum both in form and content to enable its graduates to acquire the necessary knowledge, skills and attitudes which will enable them to work with small farmers, who form the majority of Tanzanians. It was against this background that SUA took steps to incorporate the FSR/E approach in its curriculum.

3.2 The Approach to Institutionalization of Training at SUA:

Although research using a "systems" approach started at the university in the seventies with a research project on "inter-cropping" supported by IDRC, little progress has been made in introducing "systems" perspective in our training programmes. Occasionally some students have done their special project research within an FSR research project, but even then the students have tended to examine only a particular set of variables, and have hardly acquired the necessary skills or attitudes for an FSR/E work.

Likewise SUA staff have been exposed to the Farming systems approach through a series of short-term training programmes especially for those involved in research projects with a Farming Systems component. But their impact has been limited to the research projects they are engaged in, and hardly have they been able to influence the bulk of the students at SUA.

Only until recently did SUA with the support of CIMMYT take deliberate but modest steps to influence the teaching, in the University by trying to slant it towards a systems perspective.

The SUA approach involved two phases: The first phase which lasted for 3 academic years (1988 to 1990) was meant to develop awareness, both within the staff as well as the students, of the farming systems perspective in research, extension as well as in their teaching. This was achieved through several activities:

- sensitization meetings of Heads of Departments and selected members of staff, to discuss the needs and strategies of introducing the 'systems' approach in the SUA curriculum. Specific modalities and concepts to be incorporated were discussed. These were based on the CIMMYT's On-Farm Research with a Farming systems Perspective (OFR/FSP) methodology.
- Study visits by Heads of Departments and selected members of staff to research institutions and project where the approach had been successfully adopted. Study tours were conducted in kenya and Ethiopia.
- Attendance of regional training workshops on the CIMMYT methodology by selected members of staff. CIMMYT supported a number of staff to attend the regional training programme in Harare, Zimbabwe.
- Provision of reference materials on the OFR/FSP approach to the SUA Library.
- Assistance in the development of local training materials by SUA staff with support from CIMMYT.
- Organizing a national workshop of agricultural researchers and trainers to work out ways and means of institutionalizing the OFR/FSP approach in the curricula of agricultural training institutions in Tanzania.
- Incorporating "farming systems" concepts in the teaching of certain selected courses - both undergraduate and postgraduate - without necessarily altering the respective course contents. The concepts to include in each course were discussed and decided upon in a meeting of the Heads of Departments and concerned course instructors.

The first phase with covered only three Departments, Rural Economy; Crop Science; and Agricultural Education & Extension and which can be considered a pilot phase, was then followed by a second phase which, based on the experiences gained in phase I, introduced specific courses into the curricula on various 'farming systems' concepts covering research and extension.

This second phase is being implemented alongside a drastic restructuring of the whole Faculty of Agriculture curriculum, whereby instead of offering mild specializations in the B.Sc. (Agric.) degree programme, several specialized full degree programmes will be offered alongside a B.Sc. (Agric.) general programme. The B.Sc (Agric) general programme is meant to prepare mainly extensionists, while the B.Sc. (Agronomy) and B.Sc. (Animal Science) will prepare mainly researchers for crops and livestock respectively. The B.Sc. (Agricultural Economics) programme will prepare mainly policy analysts, managers, marketing and credit specialists.

However, an important feature of this whole new programme is the incorporation of Farming Systems courses cutting across all the degree programmes. These courses include:

3.3 RE 310 Introduction to Farming Systems Concepts and Research Methodology (1.5 credits)

Definition of the farming systems approaches to research and extension. The concept, objectives and evolution of farming systems approach to research and extension. Crop and livestock interaction and the role of farming systems approach in establishing a sustainable agricultural system. The definition of farming systems, technology, technological components and on-farm research. Recommendation domain. Differences between farming systems research and commodity research and their complementarity. Stages of farming systems research and extension. Meaning and purpose of the research process in the scientific enquiry. Steps in the research process- problem identification and statement, conceptual framework, developing objectives/hypotheses and research questions, reviewing pertinent literature, research designs, data collection procedures, data analysis, data presentation, research budgets and time-tabling; steps in writing a research report. Writing for publication.

3.4 HT 305 Horticultural Cropping Systems Research and Development (1 credit)

Position and functions of horticulture in farming systems. Agro-socio-economic aspects of integrating horticultural crops. Methodologies to identify constraints, potentials and priorities through farmer participative research and extension programmes in horticulture. Field case studies of approaches and strategies to horticultural crop development.

3.5 RE 404 Farming Systems Research, Extension and Development (1.5 credits)

On-farm research planning and experimentation; steps and criteria for selecting farmers for on-farm experiments; assessment of on-farm trials; agronomic, statistical and economic analysis. Partial budgets, marginal analysis, minimum return analysis and sensitivity analysis of experimental results; recommendations and diffusion of innovations, importance of gender analysis in the whole process of farming systems approach to research and extension.

3.6 Other Supporting Courses

In addition to these 'systems' courses other supporting courses have been expanded. These include Agricultural Extension, Rural Sociology, Biometry and Computer Science, which will also be taken by the majority of the students regardless of the degree specialization.

It is hoped that with this emphasis in their studies, SUA graduates will be better equipped to identify and deal with the problems of the Tanzanian farmer.

Thus, SUA chose a gradual approach incorporating Farming System concepts and methods in the curriculum, in order to ensure acceptability by all SUA staff and to prevent unnecessary hostility from those not yet sold on the idea.

SUA has also played an important part in the attempt to institutionalize Farming systems training in other agricultural training institutes in the country. In collaboration with the Farming System Research National Coordinating Unit in the Ministry of Agriculture and Livestock Development, and with CIMMYT support, SUA organized a workshop involving representatives from the various training (certificate and diploma) institutes to chart out ways of incorporating the approach in their curricula.

As a result of a workshop held in November 1990, the Ministry now has a training programme ready to be introduced into the institutes which aims at sensitizing students (most of who will be front-line extension workers and research assistants) to the concepts and methodologies involving in the Farming Systems perspective, so that as they work with senior colleagues already knowledgeable in the approach, they can understand and participate in this approach.

4. CONCLUSION: THE CHALLENGES AHEAD

Tanzania is among those countries committed to the Farming Systems approach, as evidenced by the participation of several donor organizations in supporting Farming Systems initiatives in the country. More importantly, the government has incorporated Farming systems (albeit as a separate structure) in the National Research System. Currently there is an Assistant Commissioner in charge of Farming Systems Research, who is at par with two other Assistant Commissioners - one in charge of crops research, and the other livestock research. This is a measure of the importance the government has accorded the approach.

Likewise, steps are being taken to influence the young professionals who will be coming out of the agricultural training system, towards a systems perspective of viewing the farmer situation. This is the most important step in ensuring that the approach is firmly embedded and sustained in the national systems. Indeed, according to Anandajayasekeram (1987)

"It is important to recognize that.... in-country training should be the business of the country and its higher learning institutions. Therefore, attempt should be made to convince the local institutions and administrators that institutionalizing in-country training capacity is in their best interest (pg 141).

However, as far as efforts at the University are concerned, full institutionalization will be hampered by several challenges which must be overcome:

i) The entrenched tradition among researchers and other scientists, of thinking and acting only along strict disciplinary boundaries and failure to appreciate what colleagues in other disciplines are doing. For many academics the merits of the systems approach have yet to be demonstrated.

ii) Lack of attraction for Farming systems literature. Currently only few journals are willing to publish literature on farming systems activities. Thus the rate at which literature is being generated is also slow, leading to shortage of teaching materials.

iii) The relative isolation of the university and low level of involvement in the National Agricultural Research, Extension and Training systems. There are very few areas where the University is institutionally linked to government operations. As such opportunities for staff and students to be involved and to learn from the practical realities of the country are limited.

iv) Lack of a critical mass of trainers of trainers (ToTs). Currently the few scientists versed in the approach are not necessarily well prepared to train others, who can in turn train students.

v) Inadequate training resources, particularly since a practical training in the approach will require many days of field work both for staff and students. The university does not have adequate funds, nor transport facilities to enable training to be conducted as it should.

vi) The variety of definitions, terminologies and techniques being propagated by various donor agencies and international organizations are likely to cause confusion among staff and students. Up to this moment, no steps have been taken to standardize them.

A strategy to overcome at least some of the challenges might consist of the following activities:

a) Continued sensitization of academic and technical staff, through in-house training programmes, and study visits to successful programmes, so that as many of the staff as possible are, at least, convinced of the utility of the approach.

b) Likewise developing a critical mass of ToTs, through short-term regional training programmes for keen researchers and trainers who can then spearhead training activities in the respective countries. This should involve training methodologies, training materials, and curriculum/programme development.

c) Provision of training materials, in form of case studies, relevant periodicals and other material, to be stocked in the Library.

d) Provision of resources - particularly transport - to enable staff and students to spend more time in the field,

e) A national or regional workshop to work out a standardization of terminologies, techniques etc., so that all scientists and trainers operate with the same understanding, and

f) To continue to pressure government, so that in the long run the political and resource commitments already made are strengthened, and are accompanied by gradual restructuring of the systems, so that a systems perspective pervades all the research, extension and training establishments in the country.

REFERENCES

Association for the Advancement of Agricultural Sciences in Africa (AAASA) (1975). Making Agricultural Research More Meaningful to Farmers: Proceedings of the Second General Conference of AAASA, Dakar, Senegal.

Anandajayasekeram, P. (1987). "CIMMYT's Mode of Collaboration in OFR Training for National Research Programmes", in Report of Research and Extension Administrators Networkshop on OFR Issues, held in Lilongwe, Malawi May, 1987, Harare, CIMMYT Regional Office.

Brokensha, D., D.M. Warren, and O. Werner (eds) (1980). Indigenous Knowledge Systems and Development, Lanham, University Press of America.

Collinson, M.P. (1984). "On-Farm Research with a Systems Perspective as a Link Between Farmers, Technical Research and Extension", Paper presented at the Networkshop on Extension Methods and Research/Extension Linkage, Eldoret, Kenya, June 1984.

De Gregory, T.R. (1969). Technology and the Economic Development of the Tropical African Frontier Cleveland, Case-Western Reserve University Press.

Friedrich, K.H. (1988). "Farming Systems Approach for the Effective Development of Small Farm-Household Systems: Methods and Procedures", Rome, FAO.

Huxley, P. (1976). "Curriculum Development or Disaster? A Course for Agriculture", Inter-University Council, UK. No. 23.

Keregero, K.J.B. (1981). "A Study of Identifying Critical Requirements for the Job of Extension Workers in Tanzania as a Basis for Developing a Strategy for Designing Training", Ph.D. Thesis, University of Wisconsin-Madison.

Mattee, A.Z. (1988). "Degree Level Agricultural Extension Training in Tanzania", in A.Z. Mattee et al (eds) Training for Effective Agricultural Extension in Tanzania, Morogoro, Tanzania Society of Agricultural Education and Extension.

Mattee, A.Z. (1986). "The Teaching of Agricultural Education and Extension at Sokoine University of Agriculture", in A.Z. Mattee et al (eds) Agricultural Education and Extension Training at Tertiary Level in Africa, Morogoro, Department of Agricultural Education and Extension, Sokoine University of Agriculture.

Merrill-Sands, D. (1986). "Farming Systems Research: Classification of Terms and Concepts" in Experimental Agriculture 22: 87-104.

Moris, J. (1976). "The Type of Graduates Needed to Bring About Agricultural Changes", in C.L. Keswani et al (eds) Proceedings of a Workshop on Agricultural

Curriculum for Undergraduate Studies, Morogoro, September, 1976.

Norman, D. and M.P. Collinson (1985). "Farming Systems Research in Theory and Practice", in Agricultural Systems Research for Developing Countries. Proceedings of an International Workshop, ACIAR Proceedings No. 11.

Norman, D. (1980). "The Farming systems Approach: Relevancy for the Small Farmer", East Lansing, M.S.U. Rural Development Paper No. 5.

Nyerere, J.K. (1984). "The Inauguration of Sokoine University of Agriculture", Morogoro, 26th September, 1984 (mimeo).

Opio-Odongo, J.M.A. (1986). "Farming systems Research and African Universities", The Rural Sociologist, 6(3): 206-215.

FARMING SYSTEM RESEARCH: DEVELOPMENT OF INTERDISCIPLINARY RESEARCH AT BRAWIJAYA UNIVERSITY, MALANG, INDONESIA

by

Semaoen I.[1], Liliek A.[2] and G. Zemmelink[3]

1. INTRODUCTION

This paper addresses the incorporation of a Farming Systems Research approach into an (existing) research system. In the case of this paper that is the research system of Brawijaya University (UNIBRAW), Malang, Indonesia. The first part of the paper describes the tasks and organization of UNIBRAW. The second part deals with the development of interdisciplinary research at UNIBRAW, with special emphasis on the so-called INRES project. The final part gives a brief account of thoughts about institutionalization of interdisciplinary Farming Systems Research, based on experiences at UNIBRAW.

2. INSTITUTIONAL BACKGROUND

The Brawijaya University is a State University located in Malang, East Java, Indonesia. UNIBRAW is sub-divided into nine Faculties: Medicine, Administrative Science, Law, Basic Sciences, Technical Science, Agriculture, Animal Husbandry and Fisheries. The number of students is 13.000 and the number of staff members (not including the administrative staff) 1.200. In terms of qualification of staff (number of Ph.D. holders), the Faculty of Agriculture, with 35 PhD's, is the most advanced.

In the Higher Education system of Indonesia, all staff members are required to participate in the execution of three tasks of the university, called Tri Dharma

---

[1] Brawijaya University, Malang, Indonesia

[2] Brawijaya University Malang, Indonesia

[3] Agricultural University, Wageningen, The Netherlands

Perguruan Tinggi: teaching, research and community service. The promotion system obligates all staff to obtain credit points in each of these three tasks.

Each University is free to select its Main Scientific Focus as a strategy to develop its capability to contribute to science and national development. The reason for this is that all universities are confronted with limited physical facilities, working funds and human resources. The Pattimura University at Ambon on Maluku island, for instance, opted for Marine Science and Diponegoro University, Semarang, Central Java for Coastal Village Development. For Brawijaya the last of the above mentioned three tasks (community service) is directed at Rural Development. To increase overall efficiency also research is focused on rural development problems.

2.1. Organization of Research

The organization of research at UNIBRAW follows the rules and regulations issued by the National Government (Government Law No. 30, 1990). Two types of institutions are established at the University level, one for coordination and management of research and the second for the coordination of community service. The first are referred to as Research Centre. A Research Centre is responsible for the execution of the mission of the University as reflected in its research programme. Interdisciplinary services involving different Faculties is needed to accomplish this mission. Cooperation between Departments belonging to the same Faculty is not sufficient to form a Research Centre. The Research Centres are directly responsible to the Rector. A university may have more than one Research Centre, each of them for the coordination of a particular field of research, involving more than one discipline, e.g. woman studies, watershed management, environmental studies, etc. If there are four or more such research centres within a university, they can together apply for the higher status of University Research Institute. This has to be approved by the Directorate General for Higher Education. The main function of the Research Institute is to coordinate the work of the different Research Centres.

Each Research Centre must meet, among others, the following six requirements: i.e. (1) approved mission, (2) a programme for five years, (3) sufficient human resources, (4) sufficient funds, (5) party with declared interest in the results (user), (6) commitment to publish and disseminate research results.

At this moment UNIBRAW has established five groups of researchers to form five Research Centres. i.e. (1) Centre of Environmental Studies and Resource Systems, (2) Centre of Social Sciences, (3) Centre of Population Studies, (4) Centre of Rural Industrialization, and (5) Centre of Women Studies. Based on those, UNIBRAW will apply for the status of Research Institute.

2.2. Mission

In 1977, UNIBRAW selected the mission of Rural Development as the main focus for its community service and research. Later, it was realized that this mission is very broad since it includes all aspects of life in the rural areas and all sectors of rural development. It was felt that a more specific goal had to be set. In 1991, at the beginning of the University Grand Development Plan 1991-2000, therefore, the mission was reformulated as "Development of Science and Technology in Support of Rural Industries". The term 'industries' is understood to also include agriculture. The key word remains the same, namely rural.

2.3. Function of the Research Institute

The function of University Research Institutes is to manage research throughout the University: planning, reviewing research proposals and reports, monitoring research conducted by members of the faculties, and dissemination of research results.

At UNIBRAW four sources of research funds are distinguished: (1) internal university funds, (2) research contracts, (3) the Directorate General of Higher Education (funds from various donors are channelled through the Research Management Department of DGHE), and (4) international cooperation.

UNIBRAW has adopted a one-door-policy for management and financing of research. Sixty percent of the internal funds are divided between the faculties, the other forty percent are managed by the Research Institute. The Faculties are given authority to accept research contracts, but such contracts have to be reported to the Rector through the Research Institute, and the financial management is centralized. Peer review is established at the University and Faculty level. Research proposals and reports have to be approved by peer reviewers at both levels before they can be submitted. Proposals for research to be financed by the Directorate General for Higher Education (DGHE) are submitted through the Research Institute.

3. INTERDISCIPLINARY RESEARCH AT BRAWIJAYA UNIVERSITY

The mission of UNIBRAW, Development of Science and Technology in Support of Rural Industries, obviously requires the execution of disciplinary research in various fields. It is widely recognized, however, that solving problems related to rural development requires an interdisciplinary approach. An interdisciplinary approach is also required for the UNIBRAW teaching programmes. Before they graduate, students are required to participate in the implementation of community service work. After obtaining a certain amount of credit points, fifteen to twenty students, from various disciplines are stationed in a sub-district, to help in village development projects covering all fields, such as applied nutrition, family planning, rural community development, etc. The main aim of this study-service programme is to give students a chance to learn from own experience in the village.

3.1. Early Development of Interdisciplinary Research

Interdisciplinary work at Unibraw is conducted at various Faculties. For example, the Faculty of Medicine has conducted a study to identify factors determining the high incidence of cleft-palate and other abnormalities in Nusa Tenggara Timur. This project lasted more than three years and was executed in cooperation with the Government of Australia, and it is now being proposed for funding by World Bank. Several disciplines within Faculty of Medicine participate in this project and also the UNIBRAW Faculty of Basic Sciences as well as economists, sociologists and anthropologists from the University of Indonesia, Jakarta and Airlanga University, Surabaya.

Because the subject of the Expert Consultation is Farming Systems Research the discussion below will be on the development of interdisciplinary work in agriculture.

In 1984, the Research Centre of UNIBRAW, in cooperation with the Agro-Ecosystem Research Group formed by outstanding researchers at national level, conducted a workshop with the objective to introduce agro-ecosystem analysis. This workshop was supported by the Ford Foundation and USAID, and attended by researchers from almost all agricultural research institutions in Indonesia and representatives from several universities. The focus of the workshop was on the identification of key-questions for the development of steep slope upland areas. The participants were representatives from the disciplines agronomy, tree crop science, soil science, animal husbandry, extension education, sociology, agricultural economics and remote sensing The researchers were recruited from the UNIBRAW Research Centre, the Malang Research Institute for Food Crops, and the Department of Remote Sensing of the Faculty of Geography, Gadjah Mada University, Yogyakarta.

The first phase of workshop consisted of two parts: (1) general introduction on agro-ecosystem analysis, and (2) execution of a Rapid Rural Appraisal (RRA) in the steep slope upland areas in the district Malang. The participants were divided into two multi-disciplinary groups to conduct field observations and formulate key questions for improving the agro-ecosystem. This second phase was used for more detailed observations, collection of secondary data, and group interviews. The data resulting from this were combined with those of the RRA to map the areas for classification of the steep slope agro-ecosystems. The third phase of the workshop was conducted one year after, to finalize the report.

In 1989, the Faculty of Agriculture of UNIBRAW started a study on the feasibility for expanding the area of garlic, which is now only grown by farmers in limited areas at higher altitudes. Disciplines involved in this project include agronomy, soil science, socio-economics, plant protection, etc. Similarly, the Faculty of Animal Husbandry conducts studies on dairy farms including nutrition, breeding, preventive health, economy, etc.

The latter two examples serve to show that the need for an integrated approach is recognized by the respective agricultural Faculties of UNIBRAW. Such projects are mostly funded by the Indonesian Government, in particular the Ministry of Higher Education, the Ministry of Agriculture and the Ministry of Cooperatives, i.e. do not depend on foreign funding. It is also true, however, that such projects are usually aimed at specific problems or commodities and that the interdisciplinarity is based on cooperation between specialized departments within one Faculty. Real Systems Research as first initiated by the above project on steep slope upland areas at the Brawijaya University, found its continuation in the INRES project (Interdisciplinary Agricultural Research) which started in 1989.

3.2. Background of INRES

INRES is part of the long term cooperative programme between Brawijaya University on the one hand, and the Wageningen Agricultural University and Leiden State University in the Netherlands on the other hand. This cooperative programme is supported by the Netherlands University Foundation for International Cooperation (NUFFIC). It started in the 1970's with a series of projects aimed at developing the research and teaching capabilities of specialized faculties and departments of Brawijaya University. The first projects were on agronomy, soil science and animal husbandry. Later, other projects on special fields within the Faculty of Agriculture, as well projects on fisheries, food Science and technology, social sciences, and applied basic sciences were added. Towards the end of the third phase of the cooperative programme it was concluded that the older projects in the fields of soil science, agronomy and animal husbandry would be phased out because the faculties and departments concerned had reached a satisfactory level of staff development, teaching and research capabilities. At the same time it was observed that the various projects had worked largely independently of each other. Little had been done to help establish integrated research and teaching programmes. While all projects were concerned with agriculture, each concentrated on one aspect and there was little interaction. One dealt with soils, the other one with certain crops, the third with animal husbandry, etc. At the farm, however, these aspects all come together. Rather than specialized dairy or maize farms, farms in East Java are mixed, incorporating a large variety of crops as well as animals. If the ultimate aim of agricultural research is improving the well-being of farming households, the performance of the farm as a whole becomes subject of study. It was recognized that research on the development of a single farm component may overlook constraints imposed by other components or competition with other components, and may result in a net decrease in overall productivity of the farm. Consequently, the role of the various components which constitute a farm and their interrelationships should be understood, involving various disciplines in an interactive form of research. It was therefore decided that in the next phase of the NUFFIC cooperative programme, the specialized projects would be replaced by a project of a different nature, aimed at the development of inter-disciplinary farming systems research, in short, INRES.

3.3. Objectives of INRES

INRES is a joint on-the-job training project in interdisciplinary Farming Systems Research involving various faculties/departments of the three parties: UNIBRAW, the Wageningen Agriculture University (WAU) and the State University of Leiden (RUL). The objective of the research is to arrive at development scenarios for the critical limestone area near the south coast of East Java, based on an quantitative and integrated analysis of all components of the farming household system, i.e. land and primary production, secondary production, farm and household economics and the farmer's decision making process. It is expected that the results of this research will be useful for future planning of rural development in the area, but INRES itself is not a rural development project as such. The more direct objective of the project is to develop the capabilities of Brawijaya in integrated Farming Systems Research, including the use of modern tools such as dynamic quantitative simulation models, and Interactive Multiple Goal Linear Programming, and the further development of such tools, e.g. development of dynamic models for situations of inter-cropping.

3.4. Structure of INRES

Four Faculties of UNIBRAW (Agriculture, Animal Husbandry, Economics and Public Administration) have together assigned 7 staff members holding a MSc-degree to the project. Similarly, the Wageningen Agricultural University has assigned two staff members holding the MSc-degree. Together, these staff members cover the fields of soil science, agronomy, agro-forestry, animal husbandry, agricultural economics, marketing, development economics and sociology. In principle, all these staff members are offered the opportunity to submit a PhD-thesis based on their contribution to the interdisciplinary team work. In addition, the Plan of Operation envisages that other UNIBRAW staff holding an MSc-degree will participate on a part time basis.

The research is guided by the combination of two Steering Committees: the UNIBRAW Steering Committee and the Dutch (WAU/RUL) Steering Committee. The members of the UNIBRAW Steering Committee include the Deans of the UNIBRAW Faculties involved in the project. Its chairman is also the Head of the UNIBRAW Research Centre. Although INRES is carried out at UNIBRAW, the Departments of AUW and RUL have an own interest in the project because part of the Ph.D. theses mentioned above (including those of Indonesian staff) will be prepared under their guidance and responsibility. These promoters form together the Dutch Steering Committee.

The daily management of the project is the joint responsibility of an Indonesian and Dutch Head INRES. The plan of operations stipulates that final responsibility is transferred from the Dutch to the Indonesian Head INRES about half way the project period.

4. INSTITUTIONALIZATION OF INTERDISCIPLINARY RESEARCH AT UNIBRAW

Although they are partly interrelated it may be useful to distinguish two aspects with regard to institutionalization: (1) budget and (2) personnel. With regard to budget it is observed that UNIBRAW bears all the costs of housing and salaries of its own staff involved in INRES, although a topping up of these salaries is paid by the project in order to compensate for working with the project full time, i.e. forsaking opportunities for earning additional income besides the basic salary which is inherent to the Indonesian system of remuneration. The running costs of the INRES project are now largely paid from funds provided by the Netherlands Government. Although running costs form a major part of the costs of the INRES project, the amount involved is relatively small as compared to the total research budget potentially available to Brawijaya. It seems feasible that Brawijaya itself covers such costs by diverting more of its future research budget towards interdisciplinary research. As discussed above, local research funds are allocated through the Research Centre(s)/Research Institute of UNIBRAW, on the basis of the quality of submitted research proposals. The Research Centre(s) provide therefore a permanent mechanism to stimulate interdisciplinary research in the future.

With regard to staff, it is envisaged that the INRES project will result in a group of staff representing five disciplines (soil science, agronomy, animal husbandry, economics and sociology) and four different faculties, most, if not all, with successful training at the PhD-level and, equally important, a profound experience in interdisciplinary and international teamwork. While in some other countries such highly trained staff often tend to look for other employment once training at this level is completed, past experience indicates that Indonesian University staff are faithful to their institution. Thus, it is envisaged that the INRES project will help to build a firm basis for future Interdisciplinary Farming Systems Research at the Brawijaya University. Institutionalization implies a process in which FSR becomes more impersonal, i.e. less dependent on certain individuals. Such a process starts, however, with the training of individuals, usually a small group in the initial stages. Following that, their knowledge, experience and enthusiasm has to be transferred to a larger group. In how far this part of the institutionalization process will succeed depends partly on the question whether the group trained first is given the opportunity to do so. In the case of the INRES group in Brawijaya one could think of two forms. Firstly, that they return to their respective faculties and departments and stimulate their colleagues there to participate in interdisciplinary research. From that position they could form a bridge to other faculties. The other alternative is to keep them together as a special group in the university, i.e. as a nucleus for a growing group in the future.

The plan of operations for the INRES project envisaged that apart from the full-time nucleus staff, also a considerable number of other Faculty members holding a MSc-degree would actively participate in its research programme. Until now, the number of such staff actually participating is smaller than envisaged. Reasons which could be identified are, amongst others, less motivation because they are not offered the opportunity to obtain a PhD with the INRES project, and secondly, that such staff members often occupy leading positions in faculties and departments with a heavy load of work in teaching, administration as well as departmental research. Heads of Departments and Deans of Faculties may have very plausible reasons for keeping such staff in their position. One of these is that such staff is needed at the department, not only to participate but also provide leadership for departmental research. Factors causing a certain reluctance on the side of faculties and departments to give their full support to interdisciplinary research are similar to those leading to problems in inter-institutional cooperation: each has an own task, as well as culture and perception, and prefers to keep full control over its resources in order to fulfil this task. The value of interdisciplinary research is widely recognized but participating in it does imply that the control over personnel and research funds is, to some degree, shared with others. This will only be acceptable if the partners involved are convinced that participating in interdisciplinary research programmes is beneficial for their own faculty or department.

In summary, incorporating interdisciplinary research in an existing institution requires the availability of a steering mechanism for research, including some centralization of control over funds and staff, but also a good understanding by individual partners of the common goal and interest.

IV

APPLYING FARMING SYSTEMS CONCEPTS TO TECHNOLOGY DEVELOPMENT

Overview and Synthesis

While technology is rarely depicted in graphical representations of the components of a farming system and its particular environment, it is vitally important as the force that fuels development. Over the past 20 years, efforts have been made to make the process of technology generation and application more relevant to farmers' needs and circumstances by applying a systems perspective. These efforts have been labelled in various ways (eg.FSR; R and D; FSR and D; FSR/E), causing enough confusion for CGIAR to request Deborah Merrill-Sands to attempt a standardized classification. This chapter considers the main factors affecting the application of farming systems to technology development and contains summaries of a wide range of field experience relevant to the subject.

The adoption of the farming systems approach by research institutions, has encountered mixed results. While success stories do exist, certain development agencies have ceased to promote the approach, as a reaction to the lack of a clearly definable research product arising from its adoption. This reaction is a reflection of the incomplete understanding of the concept, with a consequent application of inappropriate evaluation criteria. A farming systems perspective should change the process of research and only then can more relevant and appropriate technologies be developed.

Of even greater importance has been the lack of understanding of many physical scientists, both in western countries and developing countries where development institutions have been modelled on western structures. Two opposing schools of thought can be distinguished in this respect. The orthodox school, represented by a major proportion of researchers from the natural and physical sciences, can be classified as viewing development as the control of nature through new technology while discounting any influences arising from institutions or human values. It is characterized by a narrow, problem-solving approach geared to optimizing resource use through maximizing biological yield potential. The production environment is expected to adjust, or be adjusted, to the latest available technology.

The opposing school, which contains a much broader spectrum of disciplines including many from the social sciences, is more concerned with the application of technology to meet a range of development objectives. It is more aware of social and political factors and believes that development is for people and by people. The importance of farmers' knowledge and experience is stressed, together with the need for technology to "fit" a particular farming system.

This thinking represents a challenge to orthodox design, management and implementation of research. As such, it is dependent on effective re-orientation and training efforts, as well as the introduction of suitable reward systems. Although it has been pointed out that significant complementarities exist between the two philosophies, most attempts to introduce farming systems approaches have resulted in the creation of separate multi-discipinary units. Rather than a beneficial fusion occurring, an appendage of limited relevance has often been created.

This discouraging picture is lightened somewhat by the positive experiences outlined in the papers contained in this chapter. These also contain suggested models and mechanisms, for the successful introduction of the farming systems approach to the research and development process. Insights and observations relate to sensitization and provision of incentives for individual scientists to work in an interdisciplinary fashion. They also focus on problems of developing intra and inter-institutional linkages, including those with rural communities and NGOs. The need for fundamental reform of professional education, as a precondition for the institutionalization of farming systems in research and development, is also accorded great importance.

The paper by Jouve and Mercoiret summarizes the research and development approach of francophone institutions, such as **CIRAD**, **ORSTOM** and **INCA**. This approach tends to give more emphasis to political and social factors than the orthodox methodology employed in anglophone countries, as it seeks to achieve, "real-scale experimenting, in close collaboration with farmers, of technical, economic and social improvements of their production systems---".

The approach emphasizes farm-level feasibility and farmers' objectives, as well as the close study of production conditions and the considerations and constraints that affect farmers' strategies. It involves significant modifications of institutional linkages, including those with farmers, and seeks to replace the orthodox linear development of technology creation and dissemination with a reciprocal relationship.

Singh's paper is a detailed case study of a successful application of a farming systems approach to research in a region of India. It is based upon agro-ecosystems methodology that demands a clear understanding of farmers' problems. The approach also aims to present multiple options that can allow flexibility in technology selection, according to individual circumstances and current conditions. Policy issues requiring government support are also indentified.

The paper describes the fusion, in 1987, of two separate donor-supported programmes that eventually led to the networking of 15 national research centres and the development of common methodologies at both the micro and the meso levels. Administrative and budgetary adjustments have not yet been completed, and it is intended that the programme will be extended to encompass the whole range of agricultural support services. Nonetheless, a shift from a focus on cropping systems to farming systems has already taken place, and gender and environmental concerns are already being addressed.

STRATEGIES FOR INCORPORATING FARMING SYSTEMS RESEARCH AND DEVELOPMENT INTO INSTITUTIONAL STRUCTURES AND ORGANISATIONS

by

D. Gibbon[1]

1. INTRODUCTION

During the last 20 years Farming Systems Research concepts and approaches have made a significant impact on the development of more relevant and appropriate technologies and to the overall evolution of more sustainable agricultural systems in many developing countries. These benefits have been realised through the reorientation and action of scientists and the partial restructuring and reorganisation of agricultural institutional thinking within many national and international research systems [1]. Recent reviews of on-farm client oriented research in nine countries, (Merrill-Sands et al. 1989) and the review of USAID FSR/E Project experience (Frankenberger et al., 1989) have highlighted many important outcomes from many FSR/E programmes.

However, the overall incorporation of these ideas has been very uneven as different research and development systems have evolved at different rates, and also accepted the implications of FSR/E concepts to varying degrees. The many ISNAR reviews of research systems over the last 15 years reflect these differences.

The key issues that remain are:

(a) whether there have been real changes in the attitudes and orientation of scientists, extensionists, development workers towards their clients and their different needs,

---

[1] School of Development Studies, University of East Anglia, Norwich, UK.

(b) whether it will be possible to bring about the necessary intra- and inter-institutional changes in structure, organisation and linkages in order to facilitate more relevant R & D programmes and

(c) whether educational institutions can meet the needs of the future with more relevant training programmes.

This paper contains a synthesis of lessons drawn from a series of reviews, evaluations and periods of field work undertaken during the last 20 years in several countries in Africa and Asia. In most cases, the author has examined the problems of (or has been directly involved in) the introduction and institutionalisation of farming systems research within the context of international or national research programmes or projects with an FSR/E component. Education and training in agricultural research and development has also been a focus of the author's work [2].

2. BACKGROUND TO THE EVOLUTION OF FSR/E AND DEVELOPMENT.

Before setting out a discussion of the issues that currently face farming systems research and development, it is necessary to assume that readers have a knowledge of:

- the origins and philosophy of western agricultural science and its imposition and impact on the growth of agricultural research institutions in the West and in the developing countries[3].

- the global and local socio-political context of the Green Revolution, its various interpretations, and its aftermath [4].

- the growth and development of FSR/E approach, methods and practice over the last 15 years [5].

These sets of ideas, interest groups and developments have given rise to two paradigms of research and development:

1. There are many scientists, administrators and development workers who still hold that science and scientific method, exemplified by the transfer of "modern technology/control of nature school" will provide the answers to hunger, poverty and development. Such actors have a primary concern with narrowly defined problem solving activities, biological yield potential, optimisation of resource use, commoditisation and commercialisation and a world in which the natural and physical scientists are dominant .

2. On the other hand there are those who feel that scientific thought is but one component of dynamic social and political life systems in which human values, beliefs and political action (by many actors) all influence how technology evolves [6] However, it is important to note that not all those scientists who are currently working in farming systems research and development programmes have sympathy with this research paradigm.

An aspect of this tension and dichotomy of thought is summarised in two recent papers which explore the conflict between central source and multiple source models of technology development. (Biggs,1989, Biggs and Farrington, 1990) Continued support for the central source model comes, not only from powerful donors and interest groups, but also many agricultural scientists who still consider that the output from their work is or will have a major impact on the way in which resources are used by all farmers and the way technology develops. The greater importance of other influences, including farmer knowledge and experience and socio-political context, particularly for resource poor farmers running low external input farming systems, has been a recurring theme through this period. (see earlier work of Biggs and Clay, 1988, IDS,1979; Chambers,1983; Gibbon,1981)

The further significance of this line of thinking is that FSR/E projects and programmes could be making a much greater contribution to broader based rural development activities and to policies that can address the specific needs of disadvantaged groups. Despite the early recognition of this potential role, there is little evidence that progress has been made in this area. The way in which FSR/E has been handled within institutions may give a clue as to why it has made so little impact on development policy and programmes.

The francophone approach to agricultural systems incorporates a much stronger emphasis on a broader-based area and community approach - Recherche-developpement - and therefore seems to address these concerns in a more constructive way (Pillot, 1990).

The competing paradigms or approaches to technology development may be compatible, according to Bawden and others at the University of Western Sydney, Hawkesbury, (Bawden et al. 1985) provided that they are fitted into a hierarchy of interlinked problem solving and situation improving systems.(this has particular relevance to alternative educational approaches - see further discussion below). The main emphasis of much of the recent farmer-centred research (Chambers, et al.1989, Farrington, 1988, Rhoades and Booth, 1985) is that it is complementary, and not necessarily antagonistic to the centralist model research approach.

Given the nature characteristics of the prevailing scientific paradigm during the last 20 years, and the dominance in decision making of people trained in this philosophy, it is not surprising that FSR/E has had a difficult passage from its introduction to many institutions. both in the West and in developing countries, into which western educational models were exported many years ago.

Despite well documented successes of the contribution of FSR to the development of more appropriate agricultural production technologies [7], which frequently have occurred only in special project or institutional circumstances, and even then for a limited time period, the permanent impact of FSR ideas, methods and concepts on regional, national and international agricultural research systems has been limited.

Evaluators of FSR programmes have been slow to recognise that FSR influences the process of research rather than giving rise to a clearly definable end product. As a result it has been very difficult to identify and measure impact which may only become evident in the longer term, and even then may be interlinked with more complex institutional developments. It is also the case that many so-called FSR/E programmes are little more than modified crop or cropping systems or livestock improvement programmes which still have strong yield maximising objectives and seem to have great difficulty in developing into broader based activities that can respond to the priority needs of poorer farmer groups.

Perhaps as a result of this performance, there now appears to be something of a reaction by a number of donors and research institutions [8] to the apparent failure of FSR for not producing spectacular successes on a par with the initial changes brought about by the green revolution technologies.

The support for FSR approaches within many institutions and in National Research Programmes now seems to be less enthusiastic than it was ten years ago, even though most research systems have established FSR teams or units and have allocated resources to FSR/E activities such as interdisciplinary diagnostic surveys, farmer-researcher joint design of trials, on-farm research and joint evaluation of experiments.

3. ALTERNATIVE MECHANISMS FOR LINKING FSR/E TO BOTH SPECIFIC SYSTEMS OF RESEARCH AND TO BROADER DEVELOPMENT ACTIVITIES

The present structures and organisation of national agricultural research, extension and development systems in many developing countries are primarily based on models which reflect past and recent political history. In the cases reviewed by this author, many remnants of older, colonial and post-colonial divisions (often based on inappropriate western models) remain, with a dominant tendency to accommodate new scientific fields and new areas of endeavour by creating new departments or sections. This is because research and development institutes often exhibit extreme conservatism with regard to rethinking structures as a result of the arrival of challenging new concepts and ideas. Older, commodity interests still dominate in many institutions(and even between institutions eg. Bangladesh, Indonesia, India) as do many disciplinary divisions within them. Ministry structures also reflect older priorities, interest groups and needs which may no longer be appropriate.

However, it is also important to note that a number of exceptional research systems already had well integrated and innovative approaches that had all the characteristics of FSR/E but were not recognised as such (Biggs, 1990).

It is then not surprising that, in general, FSR/E has been treated in many instances, both in national and some international research institutes, as a new 'discipline'. (eg. ICARDA, Syria, ICRISAT, India, BARI and BAU, Bangladesh, MARIF, Indonesia, CRIG, Ghana,) Commonly, separate FSR/E Units have been created, giving the group of researchers responsibility for a full programme with departmental status. In some cases, new institutes have been established, for example in Thailand. This trend has of course been encouraged by donors because it normally suits them to support clearly identified areas of scientific endeavour. (note the current popularity of biotechnology and sustainability and the creation of new units or programmes in these areas).

The way in which agricultural research is normally planned, designed, managed, implemented and evaluated is inevitably challenged through the introduction of FSR/E concepts and methods and it is perhaps inevitable that misunderstanding, conflicts or isolation results when FSR/E is fully linked to all related activities within research institutes. One of the problems of the earlier FSR work at ICARDA was that scientists in the FSR group spent an unreasonable amount of time justifying their existence and explaining the significance of field work findings to other scientists who were not interacting in any meaningful way with farming families.

3.1. The Radical Option

A radically alternative model starts with the restructuring of ideas about the role and importance of systems thinking as fundamental to the way an institution is organised. (see Bawden and Macadam, 1988, Macadam, R. and N. Sriskandarajah, 1990; Roling,1989). This model of restructuring begins with a rethink about the nature and function of educational institutions. The underlying premise is that at the core of agricultural education should be a commitment to learning skills and processes and the essential complementarity between theory and practice. It is significant that these principles, which were at the heart of agricultural education in the UK and elsewhere more than thirty years ago (Holliday, 1959), seem to have been lost in the drive for narrower, reductionist approaches to agricultural science training.

Of the International Centres, perhaps CIP have gone the furthest in developing these ideas and in carrying through the implications into every area of work. A crucial part of this process has been the role and status of social scientists in the continuing evolution of the Centre.

Despite these exceptions, the radical option remains unlikely in the near future. The reasons for this appear to be :-

- the continuing unwillingness of scientists, extensionists, development workers, administrators and planners to understand the principles of FSR/E and D combined with a blind faith in transfer of exotic technology models.

- the failure of scientists and extensionists to develop a thorough understanding of the nature and operation of real life farming systems as a prerequisite to any research and development activity. The problems of poorer farming families are still poorly understood, there is still a marked inability to regard farmers as innovators, to value their knowledge sufficiently and to understand their values and priorities,

- the poor institutional back up given to applied research activities and the poor feedback linkages which would give such work the necessary status within research and development

3.2. Two Models of FSR Integration

3.2.1. Pakhribas Agricultural Centre

At national research institute level, Pakhribas and Lumle Agricultural Research Centres in Nepal provide examples of a high degree of integration of disciplinary contributions to both on station and on farm research. This is helped by the relatively small size of the institutions, a clear geographic mandate, autonomous planning, the strong support from Directors and long term funding from a single donor. (see Farrington and Mathema, 1990)   One mechanism developed by both these institutions and by the national programme is the shared diagnostic, implementation and evaluation activities which take place under tough physical conditions. This is known as the joint trek or the Samuhik Bhraman (Chand and Gibbon, 1990, Green and Bell, 1987, Mathema and Galt, 1988).

At Pakhribas, a deliberate decision was made not to establish a FSR unit but with the support of an FSR adviser, to create working groups, focused on clearly identified problem areas,(called research thrusts at Lumle) comprising of scientists drawn from all relevant disciplinary sections, technical staff and farmers. (Gibbon, Thapa and Rood, 1989). They consist of the disciplinary natural and social scientists who are considered to be necessary to study and contribute to the defined problem area. Such an arrangement enables basic and strategic supporting work to continue within the normal disciplinary structure, and a strong linkage is created through the members of the groups.

Figure 1. Pakhribas Agricultural Centre
Working Groups: Structure and Membership

SECTIONS	Fodder	Women	Regen. tech.	Soil fert.
Agronomy	*	*	*	*
Livestock	*	*		
Forestry	*		*	*
Veterinery		*		
Extension	*	*		
Seed Technology			*	
Socioeconomics	*	*	*	*

An important, necessary supporting mechanism is that changes need to be introduced into the review, planning and allocation of resources on an annual basis in order to give adequate emphasis to this mechanism. The key activity is the annual review and planning meeting at which such decisions can take place. A similar mechanism was also introduced into the BARI system (Biggs 1986, Gibbon, 1986)in Bangladesh.

3.2.2. Eastern India.

Larger institutions, and even national programmes have to develop other kinds of structures. For example, there is a need to formalise the position and role of FSR units, particularly when FSR/E is to be introduced into a rigidly hierarchical and deeply divided structure. eg. BARI in Bangladesh. While the danger of isolation remains, it may also be necessary to establish credibility in the early years of operation.

In a recent review of activities in Ford Foundation supported FSR projects in Eastern India (Gibbon,1990) a number of common problems were evident. [9] Many projects began with the minimum of resources, particularly staff, many were strongly crop biased and therefore had difficulty in responding to needs of farmers if these fell outside the original conceptual boundaries. An excellent example was in the Rice-based farming systems research project of Rajendra Agricultural University where the recent identification of the needs (help with pond design, construction and intensive fish technology) of landless fisherman in the **chaur**. If the project was going to help this group of people, new skills and expertise would have to be brought into the project and resources would have to be reallocated.

Attempts were made to improve the disciplinary mix by bringing in social scientists from other institutions, but this incorporation often gave rise to many practical difficulties in communication and coordination. An institutional model was proposed (Gibbon et al., 1990) to try to overcome some of these problems, to give the field team status, support and flexibility in decision making.

The model (see Figure 2.) consists of 5 components:

Farmer groups.(FGs)
These consist of any identifiable group of rural people with whom the team may interact. eg. women, cultivators, landless labourers, fishermen.

Core FSR team.
This is a small, active team of scientists, mainly working in villages. It consists, as a minimum, of an agronomist, a social scientist and a livestock scientist or other

essential disciplines as required eg. a forester) In addition, an extension-liaison person is needed and a senior scientist/coordinator to act as a link with the ASRAP. Other linkages are necessary between this team and other development teams working in the same areas.

Applied Science Support Group.(ASSG)
This is a group of active, disciplinary scientists, drawn from each major department or relevant discipline, prepared to work in the field. This group also needs linkages with other rurally based scientific staff; eg environmentalists, health workers etc.

Agricultural Systems Research Advisory Panel(ASRAP)
This consists of a panel of senior scientists (including the FSR Group coordinator). Representation from all main departments/institutes involved.

Figure 2. Institutional Structure: FSR East India

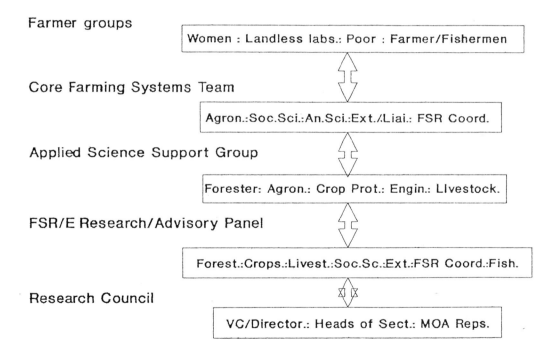

Research Council.(RC)
Normal body that reviews all research. Includes external members. Annual meeting to discuss strategy, agree on programme and reallocate resources as necessary.
Both ASRAP and RC members may also be members of Regional Planning teams who are concerned with broader planning activities at district and regional levels..

Each of these groups is given clear terms of reference and responsibilities and it is essential that all parties in the model are aware of them. Variations in the structure and balance of this model are possible. What is important is that the final structure enables and encourages the implementation of the principles of FSR/E. (Gibbon, Abrol and Singh, 1990).

3.3. Lessons on Intra-Institutional Development

1. Incorporation of FSR/E is relatively easy within small institutions when there is a high degree of autonomy, sound funding, adequate rewards and the flexibility to create alternative structures to respond to new needs.

2. Within larger institutions, FSR/E has normally been established as an attachment to existing or revised structures and as a result, full participation of actors with different, but interdependent responsibilities, is difficult.

3. Linkages with both broader and more specific areas of scientific research and development activities are often difficult unless formal linkages have been created and reward systems support interdisciplinary activities.

4. Many FSR/E programmes are still embedded in the scientist-driven western scientific paradigm. Despite much rhetoric about farmer participation and farmer-centred programmes, many only marginally involve farming families in research planning, implementation and evaluation.

5. Long term donor support is necessary in order to achieve significant progress in systems research.

4. INTER-INSTITUTIONAL LINKAGES WITH FARMER GROUPS, EXTENSION AND NGOs.

Many formal research systems are poorly linked to farmer groups, extension and to the wide range of other agencies who are engaged in developing agriculture and rural life systems. Within the recent FSR literature, there is evidence of encouraging mutual benefit from greater interaction between researchers and each of these groups of actors [10].

In reviewing recent work in this area, there do not appear to be any common approaches, except that for more effective and relevant research to develop, strong two-way linkages with all these groups are essential. The arguments would appear to be particularly strong for these linkages to be established in marginal or risk-prone areas (Wellard, Farrington and Davies, 1990). I would argue that the same principles are equally necessary in so-called 'high potential' areas where many systems are now suffering from problems arising from an overemphasis on monocropping and high external inputs. During the next 20 years farmers in many of these areas, both in the developing world and in the West (and Eastern Europe) are going to be looking for alternative strategies and techniques to diversify and stabilise their agricultural systems to include multiple community, environmental, and individual goals [11].

An important point for a research institution to consider is who, and how many, of the researcher group, should be engaged in these activities and how can time be satisfactorily allocated between them ? One great danger, arising perhaps from the 'infallibility of science' attitude, is that researchers may again tend to dominate many of these relationships, and both lose the benefit from a genuinely two-way activity, and become ineffective because they are only focusing on a very small part of a very complex system. [12]

4.1. Links with Farmer Groups

The development of interactions with farmer or user groups seems to have produced some very positive lessons for the planning and development of FSR. (Norman and Modiakgotla, 1990, Thapa, Green and Gibbon, 1988, for examples). One recurrent necessity in Asia and in Africa is that work must be carried out through the political representatives of communities. In many cases a more appropriate research interaction is with a sub-group or resource user group that has a common set of problems or focus. Recognition of such groups may not be easy, and the maintenance of working relationships with such groups may require skills that are not always found in agricultural research institutions. (Gibbon and Schultz, 1988).

Large cultural differences between researchers and potential clients remain a serious problem in forging links with groups and in developing research programmes that evolve from the real needs of those groups. This was also evident in the case from South Bihar noted in endnote 12.

4.2. Links with Extension

Linkages of research with extension systems is a major area of concern for the next decade. Significant changes in extension activities may be hastened by; the decline in the use of extension as an input supply organisation in many countries (eg Ghana and India), the limited applicability of many 'recommendations' from the centre (Nepal), the continued use of hierarchical command structures (Indonesia), separate divisions of research and extension and the mixed performance of the T and V system,(particularly in Africa) Roling ,(Roling, 1988) suggests that as both research and extension are part of the same knowledge system, they should not be organisationally separate.

Some countries, notably Zambia, seem to have tackled the problem in an interesting way by incorporating and seconding extension staff into ARPT teams (Kean and Singogo, 1988). Many countries now have extension staff engaged in adaptive field trials and demonstration pl. s which are linked to research trials. eg Bangladesh.

It would seem to be important that extension staff are much more engaged in all parts of the process of research and could take a more active, investigative role in exploring alternatives with farmers. Unfortunately, they are often trapped in a rigid hierarchical command structure, and unable to operate independently. In discussion with district extension staff in eastern Java on the potential for developing effective links with the MARIF FSR programme, the extension staff said that if they discovered any problem that required further research , or if farmers were seen to be developing unusual innovations, this information would have to be passed up through the system to Jakarta and then back through the research system to the regional research institute. (Gibbon, 1990).

A further difficulty in developing effective links between research and extension is the different educational and training background of staff. Mutual professional respect between research and extension staff was evident in relatively few cases.

4.3. Links with NGOs

Recent work reported in the ODI Network indicates the great potential for the collaboration of formal research systems with certain types of NGOs. Some of the ODI case studies indicate that researchers may only be able to develop a significant national or regional impact if they develop collaborative linkages with NGOs (Bebbington,1989; Bedergue, 1990; Gilbert, 1990; Thiele, Davies and Farrington, 1988)

In a number of countries, NGOs have been linked to the national research system for many years, for example the Mennonite Central Committee in Bangladesh (Buckland and Graham, 1990) but this has been the exception rather than the rule until recently.

The Ford Foundation supported research and development programmes in West Bengal have been developed within the Ramakhrisna Mission which is engaged in many other rural development activities, including youth club work (Gibbon, Choudhary and Chatterjee, 1990).

The joint work of Pakhribas Agricultural Centre with SNV (Netherlands voluntary agency) extension agronomists in the Mechi Hills is another good illustration of this type of collaboration. The Samuhik Bhraman are also used here to reinforce this cooperation.

This work has highlighted the essential two way nature of this kind of interaction. NGOs are becoming more numerous and more professional in many countries and many have a small number of well qualified staff. Many voluntary agencies now need greater support in information networking materials and technical backup in order to make their work more effective. Unforseen research needs are rarely built into their budgets.

In some cases, the results of interventions by NGOs create the need for research inputs which they cannot provide. An example is evident within the successful collaboration of OXFAM, ITDG and women's groups in Turkana which has resulted in a more reliable system of growing of the main food crops though improved run-off farming. This greater reliability of water holding capacity has created severe crop pest and disease problems that need human and environmentally sensitive solutions.(Martin and Gibbon 1987, Martin 1990 pers.comm.) A linkage with an appropriate research institute is essential, but there is no obvious structure or mechanism to enable this to happen.

4.4. Lessons on External Linkages

1. Good progress has been made in developing the mechanisms for working with farmer or user groups . Problems of appropriate skills training for researchers remain.

2. These kind of interactions is not yet accepted as an essential component of research practice in many research systems.

3. Mutually beneficial links with NGOs have developed in a few areas, over a long period and are only just being recognised as significant.

4. The quality of extension linkages varies considerably. No doubt that such interaction is essential, but major structural, organisational, educational and reward system problems remain. There is a longer term need for effective fusion.

5. THE NEED TO INCORPORATE SYSTEMS PERSPECTIVES INTO EDUCATIONAL PROGRAMMES AND ALSO AFFECT POLICIES AND PLANNING.

A third issue arises from the repeated re-emergence of conflict and confusion between natural scientists and social scientists, and between natural scientists with different views of science, over the purpose and need for FSR/E, and the socio-political context of science and agricultural research in particular. Much of this conflict arises from deeply held differences in philosophy of science and in the objectives and motivations of different scientists. It is clear that the education and training of agricultural scientists over the last 30 years has developed in a divergent and reductionist manner reinforced with a philosophy that supports the view that much of natural and social science is about the understanding and control of nature.(Chalmers, 1982)

While education has always responded rapidly to the challenges of 'new scientific frontiers' (eg. biotechnology), at the same time it seems to have produced scientists and extensionists who know less and less about the science and art of farming. The balance of skills and training in our education may be inappropriate as we are now are failing to evolve our methods and curricula to suit emerging needs, both of developing country farming communities, and of those who have a different set of values with regard to the use of renewable and non-renewable resources within western societies. The skills and training needed are some of very ones we discarded when we embraced the brave new world of modern, scientific, technological farming 30 years ago [13].

Many developing country agricultural universities now struggle with a structure and curriculum based on Western models developed in the 50s and 60s. These respond only slowly to innovations and new ideas from within their own countries or from contemporary western scientific knowledge. They are even more resistant to revolutionary concepts and ideas; although the one response, notably in India, Thailand and China is to create a new institution that embraces the new concept.

5.1. Orientation and Re-training

Much effort in research and development of FSR/E during the last 15 years has been directed towards retraining or reorienting conventionally trained agricultural scientists, embracing social scientists, creating multidisciplinary teams, learning from farmers, working with farmers and developing linkages with development agencies. We are also needing to carry out this kind of reorientation repeatedly which leads me to the conclusion that the basic training of many of our scientists is deficient in certain concepts and skills. This view might not be shared by many mainstream, centralist model scientists.

The creation of special courses in FSR has been one approach. Both short courses in the USA (Hildebrand,1988), the UK (Gibbon and Biggs,1985, Germany,(GTZ), France (Montpellier) and longer term courses with a strong FSR/E analytical and practical components such as the ICRA course Netherlands from 1980 onwards, from 1990 in Montpellier. MSc level courses have also been developed with a strong systems perspective in the US, the UK, Thailand and elsewhere in Asia. However, some of these ventures have suffered the same problems of isolation and lack of effective linkages with mainstream agricultural scientific education systems as the special FSR Units or divisions have done within agricultural research institutes. (Weitz, Gibbon and Pelzer 1987).

5.2. Re-thinking and Restructuring

The most radical proposal to date has emerged from ten years debate at the University of Western Sydney, Hawkesbury, where the systems philosophy has been applied to the whole way in which that education institution is organised and operates. (Bawden et al., 1985, Bawden and Macadam,1988).

The crux of this revolution is the acceptance by all - farmers, students and staff - that agricultural education is a learning system in which experiential, propositional and practical knowledge are essential components. The approach includes treatment of a range of complementary methods of analysis from soft systems to reductionist scientific enquiry. (Bawden and Macadam, 1988)

Such a premise presents a major challenge to all existing agricultural research and training institutes, as many senior academics and scientific administrators may not be willing to contemplate such a radical change. However, without such thinking and restructuring, we may need to continue to retrain scientists in the haphazard and imperfect ways that we have been doing for the foreseeable future.

Perhaps we need to return to the fundamental objectives of agricultural science training and learning and to ask for whose benefit are we making this effort ? Is it to perpetuate the dominant current scientific paradigm which ignores the fact that we live in a world in which few rural people benefit from the products of science and technology ? Or can we seriously combine our relevant knowledge and skills with those who most need to develop more secure and sustainable agricultural systems ? Can we accept that much of the debate in recent years is really about power, ethics and social justice ? These themes recur in much related writings for example, Barnett, 1988, Bawden and Macadam, 1988, Berger, 1974, Biggs and Farrington, 1990, Friere, 1972, George 1984, Rappaport 1987, Richards, 1985, Ulrich, 1988, Gibbon, 1989. This not only means empowerment to the rural poor themselves, but also to scientists and appropriate institutions to enable them to evolve organisational structures and learning methods that are appropriate to the total environments in which they operate.

5.3. Case Studies

Evidence from the case studies examined by this author has not been encouraging. Few agricultural educational institutions have grasped the nettle in the way that Hawkesbury has (even in Australia - Bawden,pers.comm.) although the same message is now being expressed by others. (see Raman,1989). Agricultural Universities seem reluctant to reform their structures, though many will take on the cosmetic changes of adding a unit or a department or a course in FSR in the same way new areas of science have been incorporated. Interdisciplinary methods of analysis and operation which include significant contributions from social scientists are rare. Relevant practical learning experiences for students have decreased at undergraduate level and often can only be gained a great personal sacrifice .

Also, despite the fact that more than 50% of the rural farming population are women and in many cases agricultural student populations contain a similar percentage of women, academic staff contain a very small number of women. (The Philippines, and to some extent, Thailand, are exceptions). Power and strategic thinking in scientific education remains in the hands of senior male academics and administrators. (Gibbon,1989)

5.4. Lessons on Incorporation of Systems Perspectives into Educational Institutions

1. There are major differences in philosophy and approach to the role of science and technology in society within our agricultural education institutions.

2. Many institutions do not appear to have evolved in response to the scientific

and technological needs now being expressed by farmers, extensionists and grass-roots community based organisations. The "products" of many institutions are further from the understanding of the needs than they were 25 years ago. A number of essential skills are absent from their portfolios.

3. Archaic and inappropriate reward systems are a major obstacle to the progress of FSR/E in many institutions.

6. POLICY AND PLANNING NEEDS.

The ultimate responsibility for making fundamental changes in direction rests with policies and planning in education and support for agricultural research and development activities. An important part of this support must be the monitoring and control of many private interest groups, both national and international. Through their activities, unacceptable negative human and environmental effects are being create for this and coming generations. We also cannot be exempted from responsibility in view of our current consumption demands.

Sustainable agricultural development can only come about through consistent and compatible policies that take into account the many and often conflicting interests of resource users. For this to happen a sensitivity to national and international political and economic forces is essential.

Those of us who are involved in the FSR/E development need to spend more time and effort in discussing these ideas with policy makers and planners - involving them in field visits, workshops and debates about the need for reform, both in the direction and balance of agricultural science and in the future education of scientific practitioners and planners.

The explosion in the growth of networking through groups in the South and North, the greater use of the media, the wide distribution of effective newsletters, eg. ILEIA, AGRECOL, ODI, Recherche Developpement, and regular regional international workshops are other ways to get these important messages across.

7. CONCLUSIONS

There is no doubt that FSR/E principles and practice are being gradually accepted into research and development systems. Many problems remain, and only a few have been touched on here. An important step forward has been the evolutionary development of the concepts and approach and the embracing of new directions and emphases. (also see Norman,1989).

Another important realisation is that addition of FSR/E to research systems is not a magic recipe for success. The early search for a quick pay off was founded on inappropriate evaluation criteria and a new set of indicators are now needed which more closely reflect the underlying objectives of FSR/E, to work with rural people in need and cope with the sustainability issue.(see Appendix 1 for criteria set derived from the author's evaluations)

If systems concepts and thinking and alternative scientific paradigms are to become integrated within our institutions and into agricultural development policies and planning, some reforms of research, training and development approaches will be necessary over the next ten years.

8. RECOMMENDATIONS

1. Research and development projects and programmes require small, multidisciplinary teams (which include farmers) that can develop a continuing understanding of farming and household systems, respond to expressed needs of farmers and interpret and utilise the experience and output of fellow scientists and other development workers.

2. Such field teams need strong back up from advisory panels and research directors and good linkages with other development agencies and activities.

3. Reward systems need to support field-based, innovatory research methods and interdisciplinary research and publication.

4. Commodity based research institutes may usefully broaden their mandates to better serve the needs of different farmer/user groups in their region.

5. Thorough annual research reviews, in which FSR/E groups, farmers, extension and other development agencies are participants, are essential. Resource re-allocation and the creation of new linkages should be possible at such meetings.

6. Greater efforts are needed to link strategic, basic or fundamental agricultural scientific research to real life systems experienced by the majority of rural farmers or resource users. Peoples' needs, ethics, sustainability and environmental sensitivity should be higher on the agenda than they appear to be.

7. Further support for the improvement of formal Research system-NGO linkages is necessary, particularly where there is a low critical number of research scientists, a large command area, many NGOs or combinations of all these.

8. More appropriate evaluative criteria are needed for research and extension activities.[14]

9. Changes are necessary in:- the way institutions respond to needs, in attitudes to poverty and empowerment issues, the openness of debate about science and society, the flexibility of structures, interdisciplinary research, learning skills, assessment methods and reward systems.

10. Training needs to have much greater interaction with the real world through the exploration of relevant practical problems.

It may be easy to be pessimistic about the chances of some of these changes taking place, given the pervasive scientific paradigm and the continuing intervention of certain vested interests, but I am reasonably optimistic that the important lessons of the past and the vast range of experience now being gathered will lead to some quite radical rethinking in many institutions.

Some restructuring of research and development institutions, and even Ministries and Departments within them, is necessary in order to achieve any of these recommendations. Tinkering with existing systems has not, so far, produced any significant change in the perceptions of the role of agricultural research institutions in agricultural and rural development.

The key might be a return to fundamental values in education and the wider acceptance of FSRD principles and concepts, not only within research institutions, but in the broader context of agricultural development.

An acceptance of a wider range of criteria, including societal, cultural and environmental goals, for the evaluation of "progress" or "development" of agricultural systems, would also be a major step forward.

Acknowledgements

This paper is partly based on several recent reports and papers by the author and also has benefitted greatly from the research and review of Samantha Bestwick, a graduate from the School of Development Studies.

Annex 1.

Criteria or indicators for evaluation of systems-based agricultural research and extension activities.

1. Does the research activity respond to national and regional research objectives and needs?

2. Is the research directed towards the needs of different socioeconomic groups of client farmers, including poorer farming families?

3. Has a farming systems approach been used, particularly in the use of interdisciplinary analysis during the initial diagnostic stage?

4. Are the findings and problems identified during initial and later survey and investigation activities channelled to appropriate research individuals, sections or institutions for further investigation or action, if necessary?

5. Are indigenous technical knowledge and farmer capabilities used in the design, development, testing and dissemination of alternative technologies?

6. Are both formal and informal research and extension methods used?

7. Are there close structural links and active interaction, both between basic, applied and adaptive research and between on-station and on-farm research activities?

8. Is the research activity well integrated with agricultural extension and other rural research and development programmes?

9. Is all the research well linked with appropriate national, regional and international programmes?

10. Is all research and extension subject to rigorous monitoring, evaluation and replanning procedures?

11. Do current institutional structures, organisational procedures, and reward systems enable and encourage research scientists to engage in many, if not all, the above types of activities?

Sources. Gibbon. Evaluation reports 1973-91.

REFERENCES:

Baker, D and D. Norman (1990) The farming Systems Research and Extension Approach to Small Farm Development. In M. Altieri and S. Hecht, eds., Agroecology and small farm development. Boca Raton, FL.: CRC Press.

Barnett, Tony. (1988) Sociology and Development. Hutchinson

Bawden,R.J, R.L.Ison, R.D.Macadam, R.G.Packham and I.Valentine.(1985) A research paradigm for systems agriculture. in Agricultural systems for developing countries. Proc. Int. Workshop. Hawkesbury.1985.

Bawden,R.J. and R.D.Macadam (1988) Towards a university for people centred development: a case history of reform. Paper for Winrock International. 1988.

Berger, P.L.(1974) Pyramids of sacrifice. Penguin Books, London.

Biggs,S.D.(1986) FSR in Bangladesh. Report for Ford Foundation,Dakha.1985. ODG,UEA.Norwich.

Biggs, S.D. (1990) A multiple source of innovation model of agricultural research and technology promotion. World Development, Vol. 18, No.11, December 1990. pp1481-1499.

Biggs,S,D. and E.Clay (1988) Generation and diffusion of agricultural technology: theories and experiences. Cont. in Ahmed,I. and V.Ruttan, Generation and diffusion of agricultural innovations. Gowers. Aldershot.

Biggs,S.D. and J.Farrington (1990) Farming systems research and the rural poor. The historical, institutional, economic and political context. Paper presented at the 10th annual FSR/E symposium, Michigan State Univ. Oct. 1990.

Berdegue, J. (1990) NGOs and farmers' organisations in research and extension in Chile. Agricultural Administration (Research and Extension) Network Paper No. 19.

Chand,S and D.Gibbon (1990) Samuhik Bhraman: a rapid and appropriate method of prioritising and replanning agricultural research in Nepal. Journal of Farming Systems Research and Extension. Vol.1.No.1 1990.

Chalmers, A.F.(1982) What is this thing called Science ? OUP UK.

Chambers,R. (1983) Rural development: putting the last first. Longman.

Chambers,R.,A.Pacey and L.R.Thrupp (1989) Farmer First. ITDG.

Checkland,P.B.(1981) Systems thinking,Systems practice. John Wiley.

Dahlberg,K.A.(1979) Beyond the green revolution. Plenum press.NY.

Farrington,J.(1988) Farmer participatory research. Editorial and six articles. Experimental Agriculture. Vol.24.3.

Farrington, J and Biggs, S.D. (1990) NGOs, agricultural research and the rural poor, Food policy 15, 16.

Farrington, J and Amanor, K. (1990) NGOs, the State and Agricultural Technology: preliminary evidence form a global review. Paper for the Asian FSR/E Symposium, Bangkok 19-22 November 1990.

Frankenburger,T., T.Finan, B.Dewalt, H.McArthur, R.Hudgens, G.Rerkasem, C.Butler Flora and N.Young. (1989) Identification of farming systems research and extension activities: a synthesis. Proc. FSR/E symposium. 1988 Fayetteville, Arkansas.

Friere, P. (1972) The pedagogy of the oppressed. Harmondsworth. Penguin.

Fukuoka, M.(1985) The natural way of farming. The theory and practice of green philosophy. Japan Publications Inc.NY.

George, S. (1984) Utopia. The university and the Third World: an imaginary cooperation programme. in ; Universities and IRd in developing countries.Ed. Bor, van den W and A.Fuller. Pudoc. Wageningen.

Gibbon,D (1981) A systems approach to resource management and development in the ECWA region: the ICARDA experience 1977-80. FAO/ECWA Conf. on Development. Damascus. 1981.

Gibbon, D. (1986) FSR in Bangladesh: a review of the BAU programme. Report for BARC. Dhaka.

Gibbon, D.(1989) University involvement in rural development programmes: some lessons from recent experience. Wageningen-Guelph Workshop on IRD, Guelph. Oct. 1989.

Gibbon, D. (1990) Selective strengthening of MARIF ATA-272 Phase V. Indonesia. Report for DGIS, The Hague. May. 1990.

Gibbon, D. B.N.Choudhury, B.N.Chatterjee, R.M.Acharya, I.P.Abrol and D.K.Singh (1990) Review of Ford Foundation supported FSR/E programme in East India. FF. New Delhi. July 1990

Gibbon, D and M.Schultz (1988) Agricultural systems in the eastern hills of Nepal: present situation and opportunities for innovative research and extension. Paper for 8th ann.symp. FSR/E. Fayetteville, Arkansas. Oct. 1988.

Gibbon, D., H.Thapa and P.Rood (1989) The development of FSR in Nepal: the working group as a focus of interdisciplinary activity. Paper for 9th ann. symp. FSR/E Fayetteville, Arkansas. Oct. 1989.

Gilbert, E. (1990) Non-governmental organisations and agricultural research: the experience of the Gambia. Agricultural Adminstration (Research and Extension) Network Paper No.12.

Hilderbrand, P. and R,Piland (1988) An inventory of short courses and university courses related to farming systems. Food Resources and Economics Department, Florida. Gainsville. Sept.1988.

Holliday, R. (1959) On the Teaching of Crop Husbandry, Agric. Progress. 34. pp 26-38.

IDS (1979) Rural development: whose knowledge counts ? IDS Bull.10. No.2.

Kean, S and L.P.Sigongo (1988) Zambia. Organisation and management of ARPT research branch. Ministry of Agric. and Water Devt. OFCOR Case study No.1 ISNAR. The Hague.

Leeuwis, C. (1989) Marginalisation misunderstood. Wageningen studies in sociology. 26. Agric. Univ. Wageningen.

Macadam, R. and N. Sriskandarajah (1990) Systems Agriculture : the Hawkesbury Approach and its implications for Farming systems Research and Extension. Asian farming Systems Research and Extension Symposium, Bangkok, Thailand, Nov. 1990

Martin, A.M. and Gibbon, D. (1987) Turkana water harvesting project review. OXFAM/ITDG, July 1987.

Matema, S and D.Gault (1988) Samuhik Bhraman: a multidisciplinary group activity to approach farmers. Paper for training course on socio-economic survey methods. Kathmandu.Nepal 1988.

Merrill-Sands,D, P.Ewell,S.Biggs and J.McAllister (1989) Issues in institutionalizing on-farm client oriented research: a review of experiences from nine national agricultural research systems. Quarterly Journal of International Agriculture, Sept. Vol.3 (4)

Norman, D (1989) Accountability: a dilemma in FSR. Culture and Agriculture. Bulletin.Spring/Summer 1989 No 38. Tuscon, Arizona.

Norman, D and E. Modiakgotla (1990) Ensuring farmer input into the rearch process within an institutional setting. AARE Network paper No. 16.ODI.Lond.

Ploeg,J.D.van der,(1990) Labour, markets and agricultural production. Westview Press.

Raman, K.V. (1989) Systems thinking in planning relevant research. ISNAR staff notes. June 1989.

Rappapaport, J.(1987) Terms of empowerment/exemplars of prevention: toward a theory for community psychology. American Journal of Community Psychology. 15. 121-145.

Rhoades, R.E. and R.H.Booth, (1985) 'Farmer-Back-to-Farmer': a model for generating acceptable agricultural technology'. Agricultural Administration, 11.pp127-37.

Richards, P. (1985) Indigenous agricultural revolution. Hutchinson. London.

Roling, N. (1988) Extension science. CUP. Camb.

Roling, N. (1989) The agricultural research-technology transfer interface: a knowledge systems perspective.ISNAR Linkage theme paper No.6.

SAD/INRA (1989) Research Units and Programmes. INRA 147, rue de l'University 75007, Paris.

Thapa, H.,T.Green and D.Gibbon (1988) Agricultural extension in the hills of Nepal: ten years of experience from PAC. AARE Network paper No.4 ODI. Lond. Dec. 1988

Thiele, G., P.Davies and J,Farrington (1988) Strength in diversity: innovation in agricultural technology development in Eastern Bolivia. AARE Network paper No.1 Dec 1988.

Trigo,E (1986) Agricultural research organisation in the developing world: diversity and evolution. Working paper No.4. ISNAR. The Hague.

Ulrich, W. (1988) Systems thinking, systems practice and practical philosophy: a programme of research. Systems Practice, 1, 137-163.

Weitz, C., D.Gibbon and K.Pelzer (1987) Farming systems development in Asia: crop/livestock/fish integration in rainfed areas. Mid-term review for UNDP/FAO. September 1987.

Welland,K, J.Farrington and P.Davies,(1990) The State, Voluntary Agencies and agricultural technology development in marginal areas. AARE Network paper 15.June 1990.

END NOTES:

1. See annual FSR/E Symposia proceedings; Kansas, Arkansas and Michigan, the Florida FSSP work, particularly the workshops, the Experimental Agriculture Series, the Systems Journals and the writings of authors who have contributed throughout the period such as Collinson, Biggs and Norman.

2. This includes; ICARDA FSR Programme 1977-80; Agricultural Research Review, Zambia 1981; Agric. Research in Somalia 1983; FSR in the SAFGRAD Programme in West Africa,1984; Agricultural Education Review, Philippines,1984; Farming Systems Research in Bangladesh, 1985, FSR at BAU, Bangladesh,1986; Farming Systems Training in Asia, 1987; Turkana Waterharvesting Project, Kenya,1987; FSR Development at Pakhribas,Eastern Nepal,1987-88; MARIF Project Review, Eastern Java,1990; FSR Workshop,CRIG,Ghana,1990; Review of FSR projects in Eastern India, 1990; FSR Annual Workshop, Tanzania, 1990; KWAP Review, Kenya, 1990; FSR Training Workshop, Mtwara, Tanzania, 1991; AFRENA Review East Africa, ICRAF, 1991. Many of the reports from this work are not published. Some are listed in the bibliography and may be available through the author.

3. see for example, Ruttan,V., Agricultural Research Policy. University of Minnesota Press, Mineapolis, 1982. Pinstrup-Andersen a. Pray, C.E., (1983), The institutional development of national agricultural research systems in South and South-West Asia.University of Minnesota. Publications of the CGIAR Network.

4. see for example; Ruttan, V.W.(1977) The Green Revolution: Seven Generalizations. IDR 1977/4. Pearse,A. (1979) Seeds of Plenty, Seeds of Want. OUP.

5. see material from; Shaner,W.W.,P.F.Philipp and W.R.Schmel,(1982) Farming Systems Research and Development, Westview Press. Gilbert,E.H.,D.W.Norman and F.E.Winch (1980) Farming Systems Research: a critical appraisal. MSU Paper 6., the output of the FSSP Project and the Annual FSR symposia proceedings, to recent work on farmer participatory research in Experimental Agriculture, Chambers et al.,(1989) Farmer First. ITDG. Also see Norman,D.(1989) Evolution and Accountability in FSR.

6. see writing of Biggs, Chambers, Farrington, Fukuoka, Richards, Altieri, Harwood, van der Ploeg, Shiva, Dahlberg, George, and many others.

7. see the proceedings of the annual FSR/E Symposia in Kansas and Arkansas and Michigan, Baker and Norman, 1990, and the growing number of papers in a wide variety of different systems journals and newsletters).

8. note the reduction of USAID support for FSR Programmes. Also, ICRISAT and ICARDA dropping of their FSR Programmes switch to Resource Management Programmes. Of the Centres, only CIP would appear to have undergone a major evolution as a result of FSR concepts and influences. Promising developments are now occurring at CIAT and ICRAF.

9. also observed in Ghana, Indonesia, Thailand (FAO Project), Somalia and Bangladesh.

10. see ISNAR, OFCOR and research-technology transfer linkage studies, FSR/E meetings 1988 and 1989, Chambers et al. 1989; Farrington and Biggs, 1990; Farrington and Amanor, 1990. and other ODI- NGO studies.

11. see work of van der Ploeg in Italy and the Netherlands,(van der Ploeg,J.D. 1990, Leeuwis in Ireland(Leeuwis,C. 1989), and SAD/INRA studies in France.

12. The University of Birsa at Ranchi provides an interesting contrast in styles. The Model Village Development Programme (strongly supported by senior staff) is presented as **the** way to develop farmer-scientist collaboration. all inputs and technologies are given to all farmers in 5 model villages, The attitude adopted is patronising ("the need to advance these backward tribal people") and the long term benefits are hard to imagine. In contrast, the modest Ford Foundation funded FSR programme includes three tribal people and a dynamic member of staff and starts with a genuine attempt to generate priorities from farmers (individuals and womens groups) own expressed needs and capabilities. This programme lacks recognition and support from senior staff.

13. In 1965 the University Grants Committee in the UK decided to cut back Agricultural education by inviting three University Agricultural Departments to phase out their agricultural degrees. The rationale for this was that the days of the general agriculturalist were over and resources were now needed to concentrate on agricultural science specialisations and stronger links with 'pure' sciences. A casualty of this policy was a Department that insisted on a year of practical farming for students and that maintained an active network of farmer collaborators who debated and worked on real farm problems together with students and staff.

14. Simplistic use of physical yield, technology adoption rates and short term economic benefits are inadequate, and indeed, can be very misleading as they fail to take account of the complexity and multiple-goal strategies of farming families or resource user groups, and also often ignore sustainability issues.(see also Lipton,M and R.Longhurst, 1989)

INSTITUTIONALIZATION OF FARMING SYSTEMS APPROACH IN AGRICULTURAL RESEARCH AND DEVELOPMENT SET UP IN EASTERN INDIA

by

V.P. Singh [1]

1. INTRODUCTION

Eastern India region comprises eastern parts of Uttar Pradesh and Madhya Pradesh and the entire states of Bihar, Orissa, West Bengal, and Assam. Rice-based farming supports about 480 million people, where approximately 80% of rice area is a wide array of rainfed agroecological conditions. Farming system performance is constrained by a complex mix of biophysical, socioeconomic and political factors. Production alternatives and input investment are restricted by recurrent flood and drought risks. Small fragmented landholdings, difficult land tenure arrangements, seasonal inadequacy of labor and farm power, unfavourable market pressure, and unconducive credit and other support services including price policies have suppressive effect in the system productivity. Lack of agroecosystem specific technologies has further weakened farmers' capacity to improve their farming system. On the whole, the slow growth in agriculture has resulted from the lack of holistic view of the complex interrelated nature of problems and needs faced by farmers.

Considering the low productivity and given the multisectoral farm enterprises pursued by the farmers, the system approach in agricultural research became imperative. It is progressively being pursued through a network of centers involving the Indian Council of Agricultural Research (ICAR), State agricultural universities, different departments in each state, and NGO's. The immediate aim is to identify farmers' problems from their farming system perspective; 2) develop appropriate technologies with multiple options to enable farmers select the most suitable ones; and 3) to pinpoint policy actions to be adopted by the government for channelling the required support.

---

[1] Associate Agronomist, Agroecology Unit, APPA Division, the International Rice Research Institute, P.O. Box 933, Manila, Philippines.

For the integrative accomplishment of these actions, operationalization of farming system approach is being pursued in the existing agricultural research and development set up through gradual "soft changes". In this context, the actual processes involved and the directions to be taken are described in this paper.

2. INSTITUTIONALIZATION IN PROCESS

2.1 Status of Research in the Past.

Prior to the initiation of the network, various agricultural research institutions in the region conducted research rather independently and usually within the confines of the research station. Each pursued commodity specific research in different disciplines within the norms and regulations of the supporting agencies. Different projects were separately operated as independent schemes without interdisciplinary approach and multisectoral consideration. Though some experiments were conducted on farmers fields, farmers participation in research process was minimal. In limited instances they were used as source of information for surveys and interviews.

The socioeconomic component, in general, was inadequately addressed for most of the institutions lacked expertise in this area; let alone attention on gender issues and other socioeconomic structural characteristics. The training component on research methodologies and system perspective was meagre. There was also the lack of communications not only among research, extension, and support services, even among and within the research teams. Research planning and evaluation at any of the institutions rarely involved all members including in special programs. In general, the farming system approach in the region was virtually non-existent.

Subsequently the above mentioned drawbacks appear to have called for holistic consideration of multisectoral problems through multidisciplinary contribution to improve the real life farming situation. The network approach adopted evidently indicate the direction set so forth.

2.2 Networking

Two networks namely, the Farming Systems Research and Extension (FSR/E) and Rainfed Rice Production were organized. There were eight participating centers in the FSR/E project (Table 1), that aimed to develop FSR methodologies through integration of biological and social scientists into interdisciplinary teams and decentralized planning for participatory research, focusing on resource poor families.

There were sixteen centers with multidisciplinary team of scientists (Table 2) in the Rainfed Rice Project. The project aimed to focus on the ecosystems-specific on-farm research for generating appropriate rice production technologies to improve the welfare of farm families. Several centers (lead and associate) were tasked for each of the rainfed lowland, upland, and deepwater ecosystems.

Of the total 18 participating centers, six were common in both the projects (Fig. 1). These projects had elements and created the basic foundation and structure needed for the operationalization of farming system concepts and the process began at different levels in some centers. However, in the initial stages the two projects worked independently (back to back) without reinforcing each other. As the leadership and the composition of teams in the two projects were different, there were duplications and difficulty of functional linkages between the projects even at common centers. Readjustments therefore, ensued before the set time frame in both the projects. The actions taken in response to these difficulties appear to be broadly based on a set of basic premises which may be listed as follows:

1. Strengthening of the linkages between the two networking line of farming system perspective.
2. Readjustment of research responsibilities.
3. Structuring of the network centers.
4. Strengthening of research skills and training capacity.
5. Participatory development of research methodologies and technologies.
6. Use of different centers in both the networks as venue for training and for developing different methodologies, in order to utilize their respective strengths and to provide equal opportunity, recognition, incentive and challenges, thereby strengthening their research and training capacity.
7. Instituting periodic monitoring and follow up of different network centers.
8. Bringing gradually other research projects in the farming system perspective.
9. Maintaining vertical and horizontal linkages with the instrumentalities of ICAR, state governments (ministries), research institutions, and extension and support service agencies.
10. Drawing of needed technical expertise from other centers of the networks and outside.

2.3 Readjustment of Research Responsibility

In the Rainfed Rice Production Project, research planning used to be centralized. As a result, many participating centers did the same set of research, irrespective of local needs, and the required level of research functions -- lead center to generate the technologies and associate centers to conduct multilocation adaptive trials.

There have been a number of adjustments in this network, however. First, the original mandate of the centers was restored in practice. Subsequently, all the centers were oriented to prepare research priorities based on their research domain specific local need, and to screen out prioritized research into those which could be dealt by the associated centers and those by the lead centers. These were discussed and finalized for each ecosystem, resulting into the adjustments in the research plans and functions of both types of centers. The research priorities of the lead center were now based on the research plans of its associate centers and their common needs, in addition to the local environment specific needs represented by the lead center, whereas, those of the associate centers were based on their local needs only. In the light of these adjustments, resources were reappropriated by the project management.

Another outcome of this exercise was the differentiation of issues to be addressed by on-station and on-farm research which involved adjustments in the composition of teams, reallocation of resources and facilities, and incentives within the center. For example, the on-farm research on the design specifications for rain water collection, conservation, and use, and on the development of suitable varieties were preceded by station research.

In addition, there emerged new issues for which modifications were made in the basic mandate of some centers. Intercropping of legumes with rice, and sequence cropping of vegetables after rice through irrigation from on-farm reservoirs and possibilities of aquaculture were identified as viable options for increasing overall farm productivity through full use of resources available in the uplands. Therefore, research on these issues was mandated as an essential component of the center exclusively for upland rice research. Similarly the research on,-- pre-rice crops, aquaculture with rice crop, and post-rice crops has also been initiated at the deepwater ecosystem centers. Such modifications are also in progress in the lowland ecosystem. The research is now directed on the whole farm system in a yearly time frame rather than seasonal studies on rice alone.

Furthermore, additional research issues and required expertise and facilities unavailable even at the so organized lead centers were identified. This necessitated restructuring of centers to pool the required services from other projects/programs of their respective institutions. There emerged the "nodal centers" from within the two networks to build up the linkages and collaboration for seeking assistance from specialized institutions from within the country and outside.

2.4 Structuring of the centers.

The multidisciplinary teams (usually 4-5 scientists), in practice earlier, were regarded as centers in these networks. In this set up, the institution as a whole was not able to participate, which often resulted into the neglect of crucial inputs from outside the team. This was also barrier to influencing the overall functioning of the institutions toward farming system perspective. Therefore, the small cell of selected scientists was expanded at two centers (Kumar Ganj and Hazaribagh) and their respective institutions, as a whole, were used as the center on a pilot basis. The new team was led by the head of the center who coordinated all the projects of that center, thereby enabling him orient the projects toward farming system perspective more effectively.

The scheme has now also become functional at Ghaghraghat, Chinsurah, Bhawanipatna, and Kendrapara centers, and is in progress at other centers. The above centers have not only been able to draw additional support from their institutions but also in gradually influencing other projects at the center. For example, the research by all scientists at Ghaghraghat are now done on dominant issues of whole farm system, identified jointly by the multidisciplinary teams of scientists, irrespective of the individual's employment under different schemes. The center also regularly draws additional support from the university staff at Kumar Ganj. Similarly, the earlier team of 5 scientists both at Bhawanipatna and Kandrapara has been expanded to 15 active members from different disciplines.

2.5 Nodal centers

Given the differences in the ecosystems and administrative setup, Central Rainfed Upland Rice Research Station (CRURRS) at Hazaribagh, Bihar, of the Central Rice Research Institute (CRRI), ICAR, and Narendra Dev University of Agriculture and Technology (NDUAT), Kumar Ganj, Faizabad, have been organized as nodal centers to represent two broad domains (Fig. 1). This has been done to strengthen both the projects, through effective use of existing research network without disrupting their functional structures.

Both network projects are ongoing in these institutions. They are also strategically located in the contiguous larger area representing the respective ecosystems: Hazaribagh in the uplands of Chotanagpur Plateau, and Kumar Ganj in the Gangetic lowland plains. Within each ecosystem, a range of environments and farming systems are found in a relatively smaller geographic area around the respective nodal centers. The other strengths of these institutions are:
- a) a number of related projects,
- b) potential for extrapolation and dissemination of methods to other institutions,
- c) training capability, facilities, and logistics, and
- d) trained manpower.

These two nodal centers have started developing need specific linkages with several national and international institutions and have been able to draw support for the network through research collaboration. Some other centers in the network, such as Raipur, Chinsurah, Pusa, and Ghaghraghat have also started shaping up like nodal centers, independently.

2.6 Training

Realizing the need for strengthening skills, especially on ecosystem analysis and FSR methodologies, and for developing the training capacity of the participating centers, trainings are organized accordingly. Given the existence of FSR activities, logistic facilities, and availability of trained manpower, different courses are organized at different participating centers, for meeting integratively the two tier training needs. Combination of different approaches are tried and refined and those effective ones are brought into practice. Training on special skills are availed at specialized institutions for selected network staff. The types of training/workshops conducted are in Table 3.

Given the training as a continuous process, its operationalization is pursued through the development of appropriate resources and through incorporating the lessons learned from the previous courses. This includes preparation of resource books, and organization of teams of core trainers. Trainings are conducted in local situations using local materials and setting. Provisions are made at the nodal centers for planning, designing, and management of training based on the assessment of needs. Efforts are being made to continue future trainings in local dialect for the junior level researchers as is done for farmers.

2.7 Research Methodologies and Their Modification

In order to fully understand the farmers' need in relation to their farming circumstances, special emphasis is placed on ecosystem analysis, prior to any FSR & D intervention. As much as possible, research methodologies are developed and adapted through the participation of all network centers and local farmers.

Ecosystem Analysis
Methodologies are developed for analyzing the dominant rice ecosystems at the micro and meso levels and are used for formulating research strategies, selecting representative on-farm research sites, and determining the extrapolation domain of technologies. Ecosystems are classified by using farmers' terminologies.

Micro Level.
The micro level ecosystem analytical procedures are developed through the modification and adaptations of available methodologies including Rapid Rural Appraisals (RRA), to differentiate the agro-ecosystems in terms of variations in

soils, hydrology, farming practices, and socioeconomic conditions. Tools used in the analyses are: spatial pattern analysis (topography, hydrology, enterprise and social class), temporal pattern analysis (process documentation of cropping, labor, food and feed availability and demand, and meteorological data), resource pattern analysis (capital and material resources), and decision pattern analysis (family decision making on technology and resource use). These tools are then pretested for their validity and reliability at selected representative key sites. Open mapping procedures are adopted and mapping scales selected.

The site analyses include a village or cluster of villages covering a cultivated area of <500 ha to more than 2,500 ha and are mapped on a scale of 1:2000 to 1:5000. Each site is classified into agroecological zones using maps, transects, flow charts, and diagrams (Figs. 2 and 3), to identify problems and opportunities in each major agroecological zone. The research problems are then prioritized using several criteria (Table 4). The priority problems are studied in detail for their biophysical causes and socioeconomic constraints affecting the application of potential solutions (Fig. 4). Possible solutions are screened prior to the conduct of experiments using different technical, socioeconomic, environmental, policy considerations.

Methodologies so developed have been effectively used by all participating centers in the region. Sixty-six sites in the Rainfed Rice Project, and 17 sites in the FSR/E project, have been analyzed and analysis is in progress at other sites. All the case studies within each ecosystem were compared and commonalities of problems and opportunities identified. This provided the empirically established generalized picture of the entire ecosystem which led to the reformulation of need-based research agenda and defining the research responsibilities of different centers. This is a major shift from the earlier centrally formulated broad program. The analysis also helped sort out the unrepresentative sites.

Meso Level.

The meso level analysis involves detailed study of a larger contiguous area, in terms of its biophysical, and socioeconomic characteristics, and delineation of the area into different homologous zones.

The main objective in developing the meso level procedures, is to evaluate the area in relation to rice and rice-based farming classification system. Various important physical, biological, and socioeconomic factors are considered and data collection procedures are pretested by teams of researchers and the extension personnel. These teams are assisted by Village Development Officers and groups of farmers, and their output compared for consistency. The procedural instruments are modified and minimum data sets prepared. The survey teams are trained in clustering, classification, and transfer of data on the base map using single mapping units at a scale of 1:25,000 to 1:100,000, and map overlays.

Delineation of different rice ecological zones in the area involves progressive overlaying of topographic, hydrologic, and climatic maps in that order. Cropping pattern, varieties used, crop management practices and input use, and selected socioeconomic conditions are separately superimposed on each of the rice ecosystem maps. The successive overlaying serves as check for validating the reliability of data as evidenced from the association between two factors in the map. Major agroecological zones are physically checked by repeated ocular inspections during field visits at selected locations and at different points of seasonal changes throughout the year. The rice ecosystems are then prioritized for research based on the extent of area, number of affected households, and potentials for research success. Field sites are identified from these maps for adaptive on-farm trials.

The results of this exercise indicate that the meso level analysis can be accomplished by the researchers as well as extension personnel provided they are adequately trained and the analysis and mapping are done with farmers' group participation. Surface water depth, crop types, and rice varieties are identified as highly reliable indicators for delineating different ecological zones. Flooding and drought patterns are also corroborated by satellite images. Physical coverage of 15-20% of the rice area is found to be sufficiently representative. The procedures so developed are being followed in Faizabad and Hazaribagh, and will be extended to other areas in the future.

Socioeconomic considerations:
Methodologies are refined and adapted for incorporating the important socioeconomic factors at different stages of research with emphasis on crucial structural characteristics (such as age and gender), and other considerations including the possible impact of the program and the technologies generated from there on employment, income distribution, and existing institutional arrangement. Interpersonal and group dynamic tools are tested for assessing the likely participation of farmers. Sensitivity, validity and reliability of field research instruments are evaluated across social classes, including the appropriate group composition of researchers, with respect to gender, for effective field studies.

By adopting farmers' local quantification units and their descriptive terminologies in relation to crop calendar by season, farmers' indigenous knowledge and practices are explored for making research germane to local conditions. Other tools and techniques include use of secondary data; field reconnaissance surveys, field observations, informal interviews, and direct field measurements; making of transects, charts, and diagrams by the farmers, and narrowly focused formal interviews.

A set of tools and techniques so developed is being tried by different centers. These activities are linked with the agro-ecosystems analysis for pursuing necessary adjustments in specifying possible solutions of the identified problems. This exercise has reportedly enhanced interaction among researchers, and farmers and researchers, and improvement in the selection of research agenda and research process. This is illustrated by two cases in Appendices 1 and 2.

More importantly, this has strengthened the feedback mechanism from farmers to the research institutions for improving the research focus, and has created a strong institutional awareness on the importance of a social science component in the farming system research. The NDUAT center has recruited one fulltime female social scientist and four fulltime female junior researchers to strengthen its on-farm research team. The university has also trained two of its extension education personnel on social science tools and methods with special focus on gender concerns, and they are included in the existing teams.

Likewise, the CRURRS at Hazaribagh center has made arrangements with the Krishi Vigyan Kendra (Extension division of ICAR) and the Holy Cross Vocational Training Institute (an NGO), both at Hazaribagh, for the participation of their female social scientists to strengthen the on-farm research team. Additionally, some of the station staff have also been trained on gender concerns and assigned to the team on fulltime basis. Arrangements are also made at the nodal centers to develop additional social science tools and methods.

The researchers of other participating centers in the two projects are trained on the social science component through training sessions in India and outside. They have started incorporating socioeconomic considerations in the research plans and in screening the technologies. Social science component at other centers is further being strengthened through additional training, seminars, and workshops.

Emphasis on Water Management.

Inadequate water supply and erratic rainfall such as the late onset of monsoon cause delays in crop establishment and drought stress at some stages of crop growth. In low-lying areas, the onset of heavy monsoon, accumulation of rain water, and slow and inadequate drainage cause delays in crop establishment and flood damages the standing crop.

Methodologies are, therefore, adapted to effectively delineate drought- and flood-prone areas and to establish their corresponding effects on farming systems performance. Indigenous farming practices in the respective situations are being evaluated for their relative strengths, and to develop the coping strategies through improving and incorporating these practices in the design of on-farm research.

The activities on the delineation of drought, and submergence-prone areas are linked with the ecosystems analysis. The available methodologies are tested at the typical sites in each of the upland, lowland, and deep water ecosystems, through field observations, repeated measurements of key factors, and discussions with the groups of farmers.

Some indigenous practices have shown clear potential for improvement and these are carefully being attended. These include:
  a) contingency provisions for arranging situational cropping systems;
  b) farm and community level rainwater collection, conservation, and use in uplands and drought-prone lowlands;
  c) utilization of shallow groundwater through low-lift pumping devices in drought-prone lowlands;
  d) minimizing flood damage in lower fields through pumping of excess water and using it for irrigating the upper fields through individual and community efforts in submergence- and drought-prone lowlands; and
  e) adjusting crop establishment schedules and methods, and tillage practices under all the three rice ecosystems.

A case on drought risk adjustment strategy by advancing the seeding date in upland is presented in Appendix 3. These methodologies are also being adopted by other centers through site visits, trainings, and pilot testing.

2.8 Improving On-Farm Research Process

The on-farm research in the region has gradually started to address the actual farm needs in the respective ecosystems and subsystems. This is brought about by various efforts mentioned earlier at different levels and by new motivations instilled among the farmers and researchers for participation. Some of the important measures being pursued for improving the on-farm research process follow next.

Farmers' Participation.
Endeavours for instituting self-propelled farmers' participation in the past have taught some crucial lessons which seem to have been used in formulating working tips.
1. Farmers dislike the prescriptive steps. They prefer to share their experiential knowledge with due recognition.
2. They are willing to participate in research if it addresses their prepotent needs, within their resource affordability and their established division of work domain. For example, while considering the tillage practices, researchers were advised to consult the 'Halwaha' (plough man). Similarly in the case of weeding, women were to be referred to.
3. They are interested only in certain components of the technology package. They are amenable to certain extent of modification in their existing system but not in the entire system.
4. Time related local farming customs are rigidly followed by farmers which require due consideration by the researchers to avoid conflict in working time between farmers and researchers. For example, starting first ploughing with bullocks and seeding are tabooed one day

after the new moon and full moon (Parva), and on the day of solar eclipse and someone's death (gami) in the community.
5. Farmers want to be involved in the entire process of research from planning to evaluation and fragmental approach dissuades them for participation.
6. Since alternative technological solutions are always compared with their local practices, it is essential to incorporate some of the local farming considerations in the actual conduct of research. Absolute control treatment in the experiments results in total failure.
7. Farmers' are reluctant to participate in complex experiments for generating multiple data. They prefer simple experiments such as comparing two treatments, where differences can be visibly compared, rather than abstract or tricky inferential.
8. Farmers prefer to conduct experiments only in part of their landholding. Research in the entire holding is seldom acceptable.
9. Farmers are consistently interested to participate whenever research is meant for raising cropping intensity and farm income.
10. Farmers' major criteria for handling research are financial, labor and time budgets. The lesser in these budgets, the more the possibility for participation.
11. The success of participation in research is highly dependent on the clear mutual understanding of the role and duties of farmers and researchers in specific activities.

All these working tips provide guidelines in the current OFR and OFT works. Refinements are in progress with corresponding research. The different levels of participatory research are being increasingly carried out through mutual coordination of the lead and associate centers.

Screening and adapting technologies.
The technologies generated by the lead centers as per the demand of associate centers are first screened by the lead centers through multilocation testing to identify the promising ones for specific environments. Those selected ones are delivered to the associate centers for further screening through multilocation on-farm trials in the different environmental conditions within the ecosystem and its subsystems. The lead centers also conduct OFT for various administrative and geographic reasons. Mutual feedback between the OFR and OFT is being channelled through seasonal meetings of all centers, reporting of results, problems and other circumstances, crop season monitoring tours, and site visits.

The associate centers conduct multilocation trials and screen the technologies for their performance in light of the problems identified earlier. The trials are periodically monitored including the ways farmers relate these to other components of their farming. This is to assess the influence of OFT in the whole farm system. Technology evaluations are done jointly by farmers and researchers and are mainly based on farmers' (participant and nonparticipant) reactions and comments to the attributes of technologies.

Handling of Technologies by farmers.

In the course of OFT, many components of the technologies are spread among farmers and there ensues their experimentation. Some aspects undergo modifications while some others are rejected. Only a few are adopted as such. To illustrate this process, three cases are presented in Appendices 1, 2, and 3. Feedback on farmers' need and preference of technologies are being used in reshaping the technology generation and transfer process.

Linking Research, Farmers, and Extension and Support Services.

In the present setting, the extension services are rendered by the state governments, universities, and NGO's. The promising technology packages from the OFT are brought by these agencies for wider dissemination through various mechanisms including farm-demonstrations, farmers fairs, training and visits, and media.

It is being increasingly realized that the extension personnel as well as support service agencies should be actively involved in the entire process of FSR & D, thereby enabling them bridge up the cleavage between farmers and researchers. The multisectoral team is likely to be more effective in interacting with farmers and catering their needs through system perspective. Likewise, it is being felt that the frequent interaction between and among farmers, researchers, extension personnel, and support service agencies would be more effective for strengthening their respective areas and gradually influencing the policy changes.

Presently the linkage between the research and farmers is relatively more developed, mainly through ecosystems analysis and OFR and OFT. But the linkages with the extension and support service agencies are minimal at all levels. At the village level it is being pursued at some centers in the network. In the prevailing circumstances the extension and support service personnel are not oriented/trained towards the FS approach, and this has impeded their participation. Alternatives exist, however, these are to be tried in selected cases and developed accordingly. The alternatives include the participation of these personnel in diagnostic, on-farm research planning, OFT and screening of technologies, and feed back stages. At higher levels, commensurate policy changes may be effected by the staff orientation on FS perspective in different sectors such as fisheries, forestry, livestock, and others.

2.9 Monitoring and Evaluation

The monitoring and evaluation (M & E) are done through five tier system. First, individual centers evaluate the process and progress of their respective teams both functional and administrative. Lead centers monitor and evaluate the research activities of their associate centers. The Director of the Rainfed Rice Project and the Coordinator of FSR/E Project evaluate the research performance of their respective centers through the Directors of Research and Team leaders of the centers. The nodal centers are evaluated by their respective Directors of Research, ICAR, Rainfed Rice Project Director and IRRI's Liaison Scientist. The overall evaluation at the eastern India level is done by a monitoring and evaluation committee appointed by the Government of India. It is composed of the Director of CRRI, Director of Research of the Nodal centers, Rainfed Rice Project Director, Coordinator of FSR/E Project, and IRRI's Liaison Scientist and Agronomist in India.

Provisions are being made for efficient monitoring and evaluation by including the administrative component and extension, support services, and farmers in the (M & E) team at different levels, beginning at the nodal centers. To reinforce the accountability and incentives among the participating scientists, an effective division of responsibility and commensurate recognition and reward schemes are being developed. The main basis considered for accountability are time bound progress, amenability for teamwork and the alike. Incentives include supportive arrangement for publication, training, and promotion.

2.10 Institutional Adjustments

The operationalization of FS concepts faced several circumstances resulting into adjustments of activities and commensurate policy actions. The focus in agricultural research is progressively being shifted from conventional approach to farming system through enhanced farmers participation, ecosystem consideration, multidisciplinary approach, and interdisciplinary action. With the development of new methodologies, integration of livestock, aquaculture, and forestry components and incorporation of gender, and environmental concerns in the regular programs have begun to take place gradually at some centers/institutions. Changes in the administrative set up, with respect to, reallocation of manpower and financial resources are also taking place. Similarly, the changes in the agricultural university curricula are being planned. The six months rural work experience requirement for graduation in agriculture has been recently implemented. Indications have emerged that the institutional changes in favour of farming system perspective are taking place at different levels in the region. However, it should be acknowledged at this point that the support from the external agencies has contributed to these changes in many ways.

3. FUTURE COURSE OF ACTION

Operational course has to be institutionalized for the overall FSR & D program within the normal agricultural research and development setup of the government. In view of the problems encountered and progresses earmarked so for, new orientations, readjustments, and improvements in the ongoing activities are deemed necessary. In addition, certain crucial areas also need to be addressed more emphatically in the future course of action. These are summarily described next.

3.1 Project integration

Further fitting in of related research projects/activities needs to be pursued according to the farmer's needs as defined for different ecosystems. Priority research should be apportioned according to the facilities and expertise of different institutions. This will involve sharing, dividing, and transfer of research projects between and among the agricultural research institutions in the region which will enhance the horizontal and vertical integration between and among them. This arrangement will also accommodate the research need of other ecosystems which co-exist geographically but are not attended to, by the ecosystem specific centers. After getting the arrangement operationalized, multi-commodity and multi-sectoral research areas can be gradually linked with the multi-level government agencies.

3.2 Human Resource Development

Whereas lateral and experiential learning is a continuous focus with pre- and in-service training backstoppings to raise and maintain researchers' skill and efficiency, the manpower in extension and other service agencies need to be continually reoriented toward FS perspective.

Special training and orientation sessions may be organized to educate the extension and credit personnel at different levels of hierarchy. The field officials should be provided structured trainings while, the executives be reoriented at right contexts. The existing training curricula of extension and credit personnel should be reviewed and efforts be made to incorporate the essence of FS perspective whenever appropriate. Discussions may be arranged with credit executives to reorient them on total credit or system credit on agriculture to cater the total household-farm credit needs of farmers. The schemes should be prepared and proposed and credit personnel be trained for trying out the scheme at pilot stage.

3.3 Agricultural Education System

Recognizing the future need of the skilled and committed manpower and professional leadership in FSR & D, efforts have to be made to develop undergraduate agricultural curricula adequately oriented toward FS perspective. Possibility can be explored on the development of some specialized graduate courses and master and doctoral students can be encouraged to undertake their thesis research with the provisions for the required research grants. Arrangement should be made with the agricultural universities in the region to reorient the six month rural work experience requirement toward the FS perspective. Prior to this, extensive discussions have to be pursued with the related authorities in the context of incorporating FS perspective in the national agricultural education system. Arrangements may be subsequently made to train and orient faculties and teachers accordingly. This can be tried at selected cases.

3.4 Farmers Organizations

Experience thus far has taught that working with individual farmers and obtaining their spontaneous participation at various stages of FSR & D has some constraints. This has somehow impeded lateral discussions and consultations among farmers to relate the OFR to inter-household need and problem structure, and eliciting technological suggestions based on farmers' interpersonal experiential perception, and the like. Moreover, organization of technology generation and delivery mechanism is being emphasized with proportionately less emphasis on the technology receiving mechanisms of farmers.

Recognizing that farmers solve many problems jointly and some indigenous innovations emerge as resultant of group efforts, attempts should be directed in the future to pursue FSR&D with farmers organizations. Small farmer groups can be organized among the participating farmers. Appropriate level of research can be allocated at different micro-ecological pockets belonging to the members of the group and research decisions be taken based on the general consensus of the group. The sense of participatory roles can be instilled to members through intra-group decisive actions.

3.5 Involving Non-Government Organizations

Although limited, past working experience with NGOs has indicated that, most of the farmer oriented NGOs are, in reality, close to farmers and are dealing with holistic problems of farming households. On their quest for credible solutions to help farmers eke out living amidst adverse circumstances they have been able to gainfully link with other institutions. Functional partnership with NGOs may foster their working relationship with farmers, thereby help progress actions both effectively and efficiently. This may be a gesture of recognition to the NGOs while attaining the concinnity in action by ruling out duplication. Arrangements may therefore, be made to strengthen the existing partnership and expand it with related NGOs toward reaching farmers through their well built course.

4. CONCLUSION

In this paper the author has attempted to draw the profile of processes and directions being followed in the agricultural research and development set up in eastern India through his working experience and available literature. It is neither a complete account of efforts of all agencies involved nor the author's lone contribution to changes. It is simply an illustration of the developments toward institutionalization of FS approach. The process as it stands is still at the experimental stage and is liable to further adjustments. The changes have not been substantial in all spheres of farming system perspective. Additional crucial areas have been identified for future course of action. Within the span of three years, there has been a series of changes in the government but the program has been able to draw consistently increasing acceptance and support from there on. The process has begun and requires continuity of efforts and policy backstoppings.

REFERENCES

TECHNICAL PROGRAM REPORTS

ICAR-IRRI Collaborative Research Program Document on the Development of Rainfed Rice Production in Eastern India. IRRI, Los Baños, Philippines. 1987.

Technical Program for 1988-89. ICAR-IRRI Collaborative Research Program on the Development of Rainfed Rice Production in Eastern India. IRRI, Los Baños, Philippines. 1988.

Technical Program for 1989-90. ICAR-IRRI Collaborative Research Program on the Development of Rainfed Rice Production in Eastern India. IRRI, Los Baños, Philippines. 1989.

Technical Program Report, (Workplan) for 1990-91. ICAR-IRRI Collaborative Research Program on the Development of Rainfed Rice Production in Eastern India. IRRI, Los Baños, Philippines. 1990.

Technical Program Report, (Workplan) for 1991. ICAR-IRRI Collaborative Research Program on the Development of Rainfed Rice Production in Eastern India. IRRI, Los Baños, Philippines. 1990.

PROGRESS/ANNUAL REPORTS

Periodical Report for 1988. ICAR-IRRI Collaborative Research Program on the Development of Rainfed Rice Production in Eastern India. IRRI, Los Baños, Philippines. 1988.

Technical Progress Report for 1988-1989. ICAR-IRRI Collaborative Research Program on the Development of Rainfed Rice Production in Eastern India. IRRI, Los

Baños, Philippines. 1989.

Technical Progress Report for 1989. Eastern India Rainfed Rice Project. 9B, Lord Sinha Road, Calcutta, India. 1990.

Summary Report of on-farm Rainfed Cropping System Research in Eastern India. ICAR-IRRI Collaborative Research Program on the Development of Rainfed Rice Production in Eastern India. IRRI, Los Baños, Philippines. 1990.

Agroecosystems characterization of Eastern India project sites. IRRI, Los Baños, Philippines. 1990.

Progress Report (1988-1990). ICAR-IRRI Collaborative on-farm Rainfed Rice Research in Eastern India. IRRI, Los Baños, Philippines. 1991.

Progress Report (1989-1991). ICAR-IRRI Collaborative Project on the Development of Ecosystem Analysis and Farming Systems Research for the Rainfed Environments in Eastern India. IRRI, Los Baños, Philippines. 1991.

Summary of Review Meeting. 1991. ICAR-IRRI Collaborative Eastern India Project. IRRI, Los Baños, Philippines.

Annual Reports for 1989 and 1990. Narendra Dev Uni. of Agriculture and Technology. Directorate of Research. NDUAT, Kumar Ganj, Faizabad, U.P., India.

Annual Reports for 1989 and 1990. Central Rainfed Upland Rice Research Station, Hazaribagh, Bihar, India.

ARTICLES

Barik, T. and B. Das. 1991. Rainfed agriculture of Kalahandi District of Orissa. In FSRE Newsletter, V. 5. Ramakrishna Mission, Lokasiksha Parishad, Narendrapur.

BardhanRoy, S.K., and S. Biswas. 1991. Agroecosystem characterization of rainfed lowland at Chinsurah Centre, West Bengal, India. Paper presented at the Planning Meeting in Rainfed Lowland Farming Systems Research. 2-6 August. Yangoon, Myanmar.

BradhanRoy, S.K. and S. Biswas. 1991. Rice-based cropping system in rainfed lowland ecosytem of West Bengal. Paper presented at the Planning Meeting in Rainfed Lowland Farming Systems Research. 2-6 August, Yangoon, Myanmar.

Biswas, S., and S.K. BradhanRoy. 1990. Rice-wheat cropping system in India.

Paper presented at the 21st meeting of the Asian Rice Farming System Working Group. 13-18 November. Hatyai, Thailand.

Biswas, S. 1990. Progress report of rice-based farming systems research in Eastern India. Paper presented at the 21st meeting of the Asian Rice Farming Systems Working Group. 13-17 November. Hatyai, Thailand.

Biswas, S. 1990. On-farm research on rice under the ICAR-IRRI Collaborative Project in Eastern India. Paper presented at the ICAR Ford Foundation Workshop on Farming Systems Research in Eastern India. India Int. Centre, New Delhi.

Biswas, S., B.P. Ghildyal and V.P. Singh. 1991. ICAR-IRRI Collaborative on-farm Rainfed Rice Research in Eastern India. IRRI, Los Baños, Philippines.

Biswas, S., R.K. Singh, S. Mallik, S. Guha, O.P. Singh, A.N. Ray, A. Kumar, and D. Panda. 1990. On-farm research to develop rainfed rice production technology in Eastern India. Poster Presentation, Int. Symp. on Sustainable Agriculture, 21st Century Asia. 19-22 November, AIT, Bangkok.

Biswas, S., S. Sankaran, and S. Palaniappan. 1991. Direct seeding practices in India. p. 55-63. In Proc. Int. Rice Res. Conf. 27-31 Aug. Seoul, Korea.

Botrall, A. 1989. The Ford Foundation supported farming systems research program in Eastern India. Paper presented in the XXIV Annual Rice Workshop. Hissar, India.

Fujisaka, J.S., K.T. Ingram, and K. Moody. 1991. Crop establishment (beusani) in Cuttack District, India. IRRI Res. Pap. Ser. No. 148. Int. Rice Res. Inst., P.O. Box 933, Manila, Philippines.

Khoshta, V.K., S.S. Baghel, and M.N. Srivastava. 1991. Physical, biological and socioeconomic characteristics and research activities of saddusankara farming systems site. Paper presented at the Planning Meeting in Rainfed Lowland Farming Systems Research. 2-6 Aug. Yangoon, Myanmar.

Lightfoot, C., N. Axinn, V.P. Singh, A. Bottral and G. Conway. 1989. Training Resource book for agroecosystem mapping. IRRI/Ford Foundation Pub. IRRI, Los Baños, Philippines.

Lightfoot, C., V.P. Singh, T.R. Paris., P. Mishra and A. Salman. 1989. Training resource book for farming systems diagnosis. IRRI/ICLARM Pub. IRRI, Los Baños, Philippines.

Lightfoot, C., N. Axinn, K.C. John, R.K. Singh, D.P. Garrity, V.P. Singh, P. Mishra and A. Salman. 1991. Training resource book for participatory experimental design. NDUAT/IRRI/ICLARM Pub. IRRI, Los Baños, Philippines.

Singh, C.V., M. Varier, and V.S. Chauhan. 1990. Farmers' wisdom in the management of rainfed rice in Eastern India. p. 128. Int. Symp. Natural Res. Management for Sustainable Agric. Abstracts. New Delhi, India.

Singh, C.V., R.K. Singh, R.K. Tomar, V.S. Chauhan, and M. Varier. 1990. Rice-based intercropping system for rainfed upland conditions of Chotanagpur plateau. IRRN 15(3):36.

Singh, H.P., R.N. Singh, L.M. Jaiswal, and N. Malik. 1991. A case for client-based research for varied rainfed shallow water depth rice ecosystem in Eastern India. In FSRE Newsletter. Ramkrishna Mission Lokasikha Parishad, Narendrapur.

Singh, P., O.P. Singh, and V.K. Singh. 1990. Risks and uncertainties in paddy production with special reference to deepwater in Eastern U.P. - a case study. p. 16 In Proc. Int. Conf. Ext. Strategies for minimizing risk in rainfed agric. 6-9 april. New Delhi, India.

Singh, R.K., V.S. Chauhan, C.V. Singh, M. Varier, D.K. Paul, and K. Prasad. 1990. Agroecosystem analysis of rainfed upland key site for on-farm research. Int. Symp. Rice Res. - New Frontiers. 15-18 Nov. DRR, Hyderabad, India.

Singh, V.P., R.K. Singh, R.K. Singh, and V.S. Chauhan. 1990. Developing integrated crop-aqua-livestock farming system for rainfed uplands in Eastern India. AFSRE sympo. Sustainable Farming System Abstracts - 21st Centry Asia. 19-20 Nov., AIT, Bangkok, Thailand.

Singh, V.P. 1991. Rainfed lowland rice-based farming systems in Eastern India: Status and research thrust. Paper presented in the Rainfed Lowland Farming Systems Research Planning Meeting, Yangoon, Myanmar, 2-6 August.

Singh, V.P., R.K. Singh, V.S. Chauhan, C. Siota and A. Frio. 1991. Training Resource book on the design of on-farm experiments. CRURRS/IRRI Pub. (In Press).

Table 1. Network centers in the Farming Systems Research and Extension Project.

CENTER	LOCATION
Narendra Dev University of Agriculture and Technology (NDUAT)	Kumar Ganj, Faizabad, U.P.
Rajendra Agricultural University (RAU)	Pusa, Bihar
Birsa Agricultural University (BAU)	Ranchi, Bihar
Orissa University of Agriculture and Technology (OUAT)	Bhubaneshwar, Orissa
University of Kalyani	Kalyani, West Bengal
Central Rice Research Institute (CRRI)	Cuttack, Orissa
L.N. Mishra Institute of Economic Development and Social Change (LNMI)	Patna, Bihar
Rama Krishna Mission (RKM)	Narendra Pur, West Bengal

Table 2. Network centers in the Rainfed Rice Production Project.

Research Institution	Center Location	Target Environment
Rainfed Lowland Ecosystem		
Assam Agricultural University (AAU)	Regional Agril. Research Station, Titabar, Assam (Lead Center)	Brahmaputra flood plains sandy to loamy soils, acidic, fertile, flood-prone
State Govt. Research Station	Rice Research Station, Chinsurah, Hoogly, West Bengal (Lead Center)	Ganga flood plains Sandy loam soils, neutral to slightly alkaline reaction, medium to high OM content and CEC, quite fertile, submergence-prone
Narendra Dev Uni. of Agri & Technology (NDUAT)	Crop Research Station, Masodha, Faizabad Uttar Pradesh (Associate Center)	Ganga Alluvial Plains, clay loam soil, neutral to saline reaction, meduim fertility, submergence and drought-prone
Indira Gandhi Agricultural University (IGAU)	Raipur, Madhya Pradesh (Associate Center)	Drought-prone
Rajendra Agricultural University (RAU)	Agril. Research Center/ Rice Research Station, Mithapur, Patna, Bihar (Associate Center)	Ganga alluvial plains, neutral soils with medium fertility and submergence prone
Orissa University of Agriculture and Technology (OUAT)	Jute Research Station, Kendrapara, Cuttack, Orissa (Associate Center)	Mahanadi Basin-flood prone
Upland Ecosystem		
ICAR-Central Rainfed Upland Rice Research Station (CRURRS)	Hazaribagh Bihar (Lead Center)	Plateau Region of Bihar - rolling topography with red lateritic soils, poor fertility, high acidity, shallow soil depth, drought prone
Bisra Agricultural University (BAU)	Kanke, Ranchi, Bihar (Associate Center)	Intermittent drought, low fertility

Table 2. Cont'd.

Research Institution	Center Location	Target Environment
Orissa Univ. of Agr & Tech. (OAUT)	Regional Research Station Bhawanipatna, Kalahandi, Orissa (Associate Center)	Plateau region of Orissa - Red lateritic soils, low fertility, black soils, less fertile, intermittent drought.
Narendra Dev Univ. of Agri & Tech (NDUAT)	Crop Research Station Kumarganj, Faizabad, Uttar pradesh (Associate Center)	Ganga plains - sandy loam neutral to slightly alkaline, and quite fertile soils, favorable to drought prone
Assam Agril Univ. (AAU)	Regional Agricultural Research Station, Shillongani, Nagaon, Assam (Associate Center)	Brahraputra flood plains - sandy to loamy acid soil flat plain or light sloppy basin, favorable.

Deepwater Ecosystem

Research Institution	Center Location	Target Environment
Narendra Dev Univ. of Agri. & Tech. (NDUAT)	Crop Research Station Ghaghraghat, Bahraich. Uttar Pradesh (Lead Center)	Deep and very deep water areas of U.P.
State Govt. Research Station	Rice Research Station Chinsurah, Hoogly, West Bengal	Deep and very deep water areas of W.B.
Rice Research Station Rajendra Agr. Univ (RAU)	Rice Research Station Biraul, Darbhanga, Bihar (Associate Center)	Deep and very deep water areas of Bihar
Assam Agril. Univ. (AAU)	Regional Agril. Research Center North Lakhimpur, Assam (Associate Center)	Deep and very deep water areas of Assam

Across Ecosystems

Research Institution	Center Location	Target Environment
L.N. Mishra Institute of Economic Development and and Social Change	Patna, Bihar	All ecosystems

Table 3. Training/workshops conducted in the two networks.

COURSE	VENUE	NUMBER OF PARTICIPANTS
In Country:		
o Agroecosystem mapping	RAU, Bihar	48
o Systems problem diagnosis	BAU, Bihar	30
o Participatory design of on-farm experiments	NDUAT, U.P.	43
o Farming systems research methodology	CRURRS, Bihar	30
o Gender sensitivity analysis in FSR	CRURRS. Bihar	23
o Rapid diagnostic survey I	NDUAT, U.P.	22
o Rapid diagnostic survey II	CRRI, Orissa	14
o Hands-on training in computer skills and data analysis	RKM, West Bengal	12
o Poster preparation and presentation	RKM, West Bengal	20
o Design, layout and interpretation of on-farm experiments	RKM, West Bengal	15
o Rice-fish farming systems	CIFA, Orissa	22
Sub-Total		279
Outside India		
o Farming systems research and extension skills	IRRI, Phil.	4
o Gender issues in FSR	BRRI, Bangladesh	8
o Lowland rice farming systems	Yangon, Myanmar	3
o Deepwater rice farming systems	Bangkok, Thailand	3
o Crop-livestock systems	Dhaka, Bangladesh	1
o Extrapolation methodologies	IRRI, Phil.	1
o Farm machinery and tools	IRRI, Phil.	1
o Water management	IRRI, Phil.	2
Sub-Total		23
Total		302

Table 4. Prioritization of Meru village problems, Hazaribagh, Bihar.

Problems	Extent	Ranking Criteria [1]			
		Frequency	Severity	Importance	Overall Rating (cumulative)
Moisture stress	9	8	9	9	8.80
Brown spots in rice	7	5	6	6	6.00
Blast disease in rice	6	4	8	7	6.25
High weed population	9	9	7	9	8.50

[1] On 1-10 rating scale; 1 least to 10 most.

APPENDIX 1

Weeding Research
(A Case Study)

In view of the pressing weed problem in the uplands, an on-farm experiment was conducted by row seeding of a high early vigor rice variety (RR167-982) at Hario village of Hazaribagh. It was expected to suppress weed population and make weeding easier in rows, thereby minimizing weeding drudgery for women. This decision was made in consultation with the male members of the household and on the basis of favorable on-station results.

The experiment was laid out with four replications in the village. In spite of providing hand hoe, the women did not weed the fields in 3 out of 4 replications almost up to 6 weeks after sowing. Therefore, the researchers weeded the field to collect the experimental data and as expected this treatment provided the best yields.

The reason for non-weeding given by males was that the females were busy in some other operations, but the females reasoned out differently. They used to weed the rice field after it was lightly harrowed (tevai) by the male members, so that they could easily pull the weeds manually. In the new system, the weeding involved hoeing as an additional burden to the women, which required more time and exertion. Moreover, interrow space was heavily covered by weeds that belied the promised potential benefit of the technology itself. Upon further probing, the women had this to say, ".... Despite knowing that weeding is our work, you did not consult us; had you consulted us in the beginning we would not let you conduct this experiment.... The RR982 is a good variety and in the future we will consider this in our tevai system of rice cropping......"

In the same experiment, realizing the higher rice yields from another variety, RR165-160, the males were eager to adopt it. However, the females were concerned about its low straw yield due to its short stature and grazing problems due to its longer maturity duration than the local varieties. It would raise their burden of fodder collection.

It seems that the new weeding technique replaced the plough with hoe and animal power by women. Also, the new variety replaced the traditional one and increased the women work burden. These changes disrupted the social interaction in weeding and crop care, thus, the technologies were socially not compatible, though economically viable.

Appendix 2

**Intercropping Research
(A Case Study)**

Intercropping technology of pigeon pea with rice was found to be productively promising for uplands in the on-station research at Hazaribagh. It provided 0.4 t/ha of pigeon pea grain yield and increased cropping intensity by 36%. The 0.6 t/ha decrease in rice yield was more than compensated by the high-value pigeon pea, and this technology provided additional income of $70.00/ha.

It was considered best suited for drought-prone conditions with built-in compensation provision for rice crop failures by the high-value pulse. Therefore, several on-farm trials were conducted to introduce this technology in Nagwan village covering 5 ha in different parcels. The trial crop performed at par with the on-station results. Impressed with its performance, farmers tried for a year but abandoned the practice the following year.

Underlying reasons given by the farmers were that there was further decrease in rice yield than the pigeon pea could compensate for. They did not consider the pigeon pea yield as a compensation for the loss in rice yield, as rice was socially more valued than pigeon pea. Moreover, the pigeon pea remained in the field longer after rice harvest and was destroyed by animal grazing. Conventionally, the post-rice fields were meant for grazing throughout the Hazaribagh uplands.

However, farmers suggested that should they be provided with short-duration pigeon pea variety maturing together with rice, they would still try out the intercropping technology with wider row spacing. In case short-duration varieties of pigeon pea are unavailable, some of them suggested to plant pigeon pea earlier than rice to match up their maturity. They also suggested that if rice + pigeon pea are planted in a contiguous area it would protect the crop from animal grazing.

Subsequent research on rice + pigeon pea intercrop indicated increased blast incidence in rice brought about by increased humidity due to pigeon pea stands. The blast disease incidence was higher in the village than at the station. Failure in specifying to farmers the rice variety used in the experiment and in thoroughly checking the associated disease incidence partly contributed to this.

The farmers' suggestion of using wider row spacing seems quite logical as it would decrease the build up of humidity. It appeared that farmers' reluctance to adopt the new technology was their concern on the additional cash and labor requirement for rice blast control.

Appendix 3

Advancing the Planting Date of Upland Rice (A Case Study)

Upland rice in the Chotanagpur plateau is planted after the onset of monsoon rains. Third week of June is considered to be the normal planting dates. Crop planted after this period is likely to suffer from drought, and insect and disease pests. Analysis of the environmental characteristics and farmers' practices showed that about two weeks time, after the onset of monsoon, is lost in the land preparation and seeding, and so with it the rain water. Farmers also encounter problems in sowing rice as monsoon starts with continuous heavy rains that delay the land preparation. The rain water could be effectively utilized by the rice crop if its seeding date can be advanced before third week of June. Farmers in this region practice advance seeding in the midlands.

On-station and field experiments were therefore conducted advancing the seeding dates by two weeks in the uplands and compared with the normal seeding dates. The 2 years' results have provided an average yield increase by 20-25% over normal dates, as the crops matured before the third week of September. This has also resulted in increased cropping intensity because a subsequent crop of niger and mustard was possible with residual moisture. In other cases, this provided sufficient time for land preparation where a wheat crop was being tested.

Some of the disadvantages associated with advance seeding of rice were the lack of synchronous emergence and heavy weed infestation. This required integrating weed management research with the seeding dates.

188 INSTITUTIONALIZATION OF A FARMING SYSTEMS APPROACH TO DEVELOPMENT

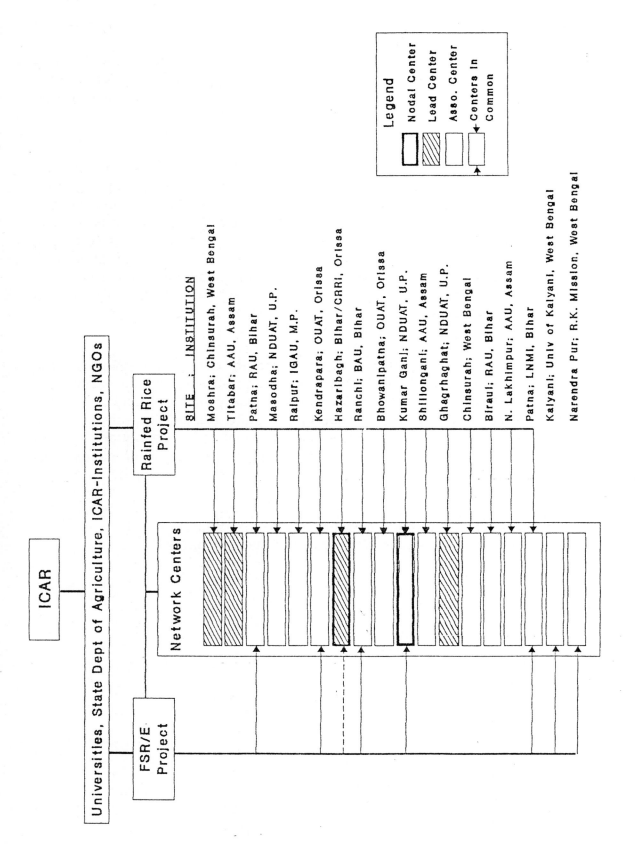

Fig 1. Network Centers in Eastern India

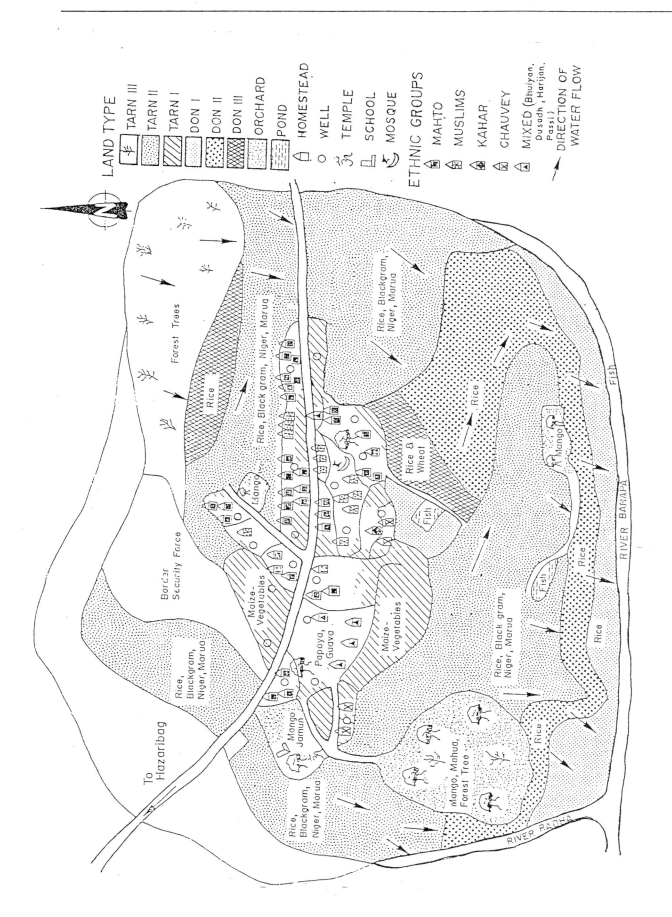

Fig. 2. Agroecological map of Village Meru, Hazaribag, Bihar, India.

LAND TYPE	FOREST LAND Non Agricultural	UPLAND		MEDIUM LAND	LOW LAND
		TANR II, III	TANR I	DON II, III	DON I
SOIL TYPE	Stoney	Sandy loam	Loam	Loam	Clay loam
FERTILITY STATUS (O.M., N, P, K, & Ca)	Low	Low	Low	Low	Medium
LAND USE	Forest/ Grazing	Forest, Agricultural, Grazing	Agricultural	Agricultural, Grazing	Agricultural, Grazing
CROPS	None	Rice, Blackgram, Millets, Niger	Maize, Tomato, Brinjal, Radish, Peas, Onion, Fenugreek, Spinach, Chillies, Cauliflower, Potato, Beans, Carrot, Garlic, Coriander, Ladies finger, Cucurbits	Rice, Wheat	Rice
WATER RESOURCE	Rainfed	Rainfed	Irrigated with wells, Rainfed	Rainfed	Rainfed
TREES	Forest trees	Mango, Mahua, Blackberry	Mango, Papaya, Bamboo, Drumstick, Guava, Jackfruit, Banana	None	None
LIVESTOCK	None	None	Cows, Bullocks, Pigs, Buffaloes, Goats, Fowl	None	None
PROBLEMS	Various (not studied)	Moisture stress, Lack of irrigation, White ants, Brown spots and blast in rice, Poor fertility, Soil erosion, Cattle grazing, Paucity of quality seeds of HYV's, High weed population	Moisture stress, Lack of labor & irrigation, Lack of improved varieties, High weed population, and diseases & insects in vegetables	Moisture stress, Lack of irrigation, Paucity of HYV seeds, Poor fertility, Blast and insects in rice, High weed population	Moisture stress, Paucity of HYV seeds, Low fertility
OPPORTUNITIES	Tree crops	1. Improved cropping intensity 2. Improved forage species 3. Short duration improved rices 4. Intercropping / Sequence cropping 5. Land consolidation 6. Composting and green manuring 7. Timely supply of seeds 8. Ridge and hedge row methods	1. Moisture conservation practices 2. High quality seeds 3. Better breeds of livestock & poultry 4. Suitable varieties 5. Land consolidation 6. Improved pest management	1. Improved varieties 2. Green manuring 3. Composting 4. Improved cropping intensity/ patterns 5. Land consolidation 6. Improved pest management	1. Improved varieties 2. Green manuring 3. Composting 4. Land consolidation 5. Cropping pattern 6. Suitable varieties 7. Improved cropping intensity/ pattern

FIG. 3. TRANSECT OF VILLAGE MERU, HAZARIBAG, BIHAR, INDIA

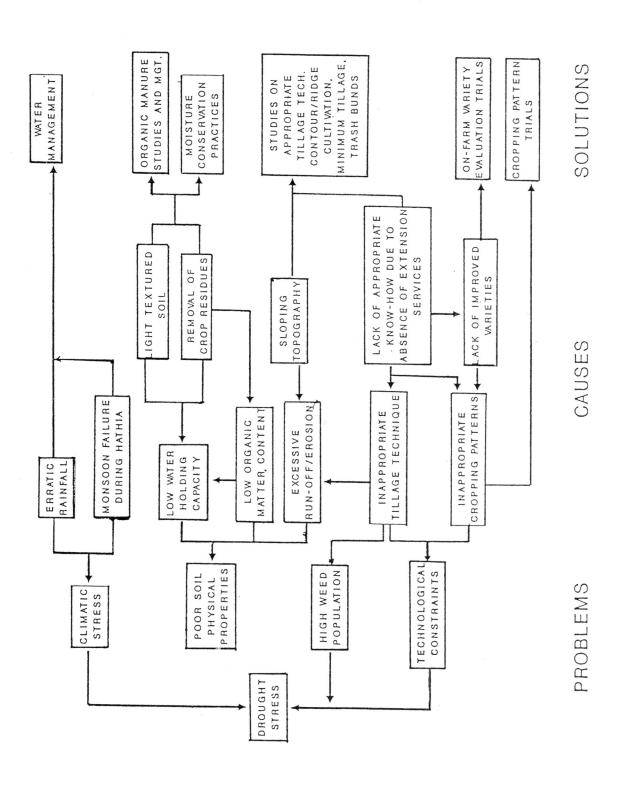

Fig 4. Problem identification, diagnosis and proposed solutions, village Meru, Hazaribag.

RESEARCH AND DEVELOPMENT: FARMING SYSTEMS RESEARCH AT THE SERVICE OF RURAL DEVELOPMENT

by

Jouve P. and M.R. Mercoiret[1]

1. INTRODUCTION

Research on farming systems has increased considerably over the past decade. It has stemmed from recognition of the difficulties encountered in the large-scale extension of technical innovations developed by agricultural research, and in particular the difficulty of adoption of such innovations by small farmers.

This new trend has stimulated many types of research, which have been the subject of a very interesting attempt at classification on the part of the Deborah Merrill-Sands at the request of CGIAR. The present communication is centred on what she called "Farming Systems Research and Development" (FSRAD) and which aims at combining the skills and responsibilities of researchers, farmers and development officers within the framework of what are referred to as "research and development operations or programmes.

Although the idea of Research and Development represents a comparatively well-defined activity in industry, its application in agriculture is much less clear. An attempt is made here to define such application in the light of experiments carried out or managed by French research teams in various parts of the world.

Research and Development can be defined as the "real-scale experimenting, in close collaboration with farmers, of technical, economic and social improvements of their production systems and the techniques used to exploit their environment". It would seem useful to begin by specifying certain basic features of this experimental work, if only to differentiate it from other research on farming systems.

---

[1] Agronomist and sociologist, Agrarian Systems Department, CIRAD, Montpellier, France

First of all, this approach is based on a simple observation: "It is not enough to test and perfect the techniques and technology which can improve farm operation and productivity; it must also ensure that the farmers who manage these farms possess the means for adopting these improvements and are interested in doing so."

It is the farmers and their ability to change, to use innovations and master their implications who finally determine whether development operations succeed or fail. Thus, a lasting process of innovation must begin with the real conditions of agricultural production and take into account the constraints and variables which affect farmers' strategies.

In addition, the Research and development procedure involves significant modification of relations between research, development bodies and farmers.

Transfer of technology has long been organised in a linear manner in which the role of research is to design innovations for subsequent transfer to development structures which in turn attempt to pass them on to farmers through extension operations. Research and Development proposes to replace this linear pattern by a triangular, reciprocal relationship between the various categories involved during all phases of the process of transformation of farming conditions.

The durable acquisition of technical innovations and, above all, mastery of these innovations by farmers depend on numerous conditions which may be associated with any of the following:

.   supply of inputs,
.   marketing of produce (prices and markets),
.   the management of farms and of natural resources.

Unlike certain types of research on farming systems, Research and Development considers that these conditions are not only external constraints but also factors that it will be sought to change; this is done in particular by collaborating with farmers to seek the type of organisation that will give them the best possible mastery of these production conditions.

Consequently, both development strategies and the conditions of organisation of farmers are studies in the same way as the functioning or improvement of farming systems. In other words, it is considered that technical innovation and social innovation are complementary, indissociable features of the same process of change and that they should both receive attention.

Finally, as soon as Research and Development takes into account the conditions of farmers' social organisation in the management of their resources, it cannot limit analysis and actions to production units alone. Its field of investigation broadens to cover the higher levels of organisation of which these production units form part; this leads, among other things, to considering methods of management of space by rural communities.

2. PROCEDURE

As in many other types of research, the procedure of Research and Development comprises three major components between which there are numerous iterations. The components are:

- analysis and diagnosis,
- experimentation of innovations,
- extension and transfer.

The specific character of Research and Development does not therefore lie in the general organisation of procedure but more in the way in which each of the phases above is applied. Since a large amount of literature has been published on the subject, this contribution is limited to the most typical and most original work undertaken in this field by CIRAD, ORSTOM and INRA research teams.

2.1 Analysis and Diagnosis

Although every Research and Development operation starts with a phase of analysis of the initial situation, it is important to note that this analysis should be continued throughout the operation so that the effects of the innovations experimented can be judged.

The result of this analysis phase is a diagnosis of the constraints and potential for improvement of the farming systems and agrarian systems with a view to determining the priority operations to be carried out.

From this phase onwards, the fixing of priorities must be the subject of concertation with farmers on the basis of presentation and discussion of the initial diagnosis. This presentation enables comparison of the points of view of researchers and farmers and reveals the major preoccupations of the latter.

The various work carried out on diagnosis shows that although analysis of the base data on the physical and human conditions of exploitation of the environment is a necessary preliminary, it is not enough. It should be complemented by analysis of methods of exploitation of the environment through analysis of farming practice. As was stressed by Pierre Milleville, this analysis of farming practice makes it possible to reveal various levels of decision, whether in crop and livestock management or in land resource management. These levels determine the various land or management units taken into account in this analysis phase and which range from region through village and individual farm to the field.

Systemic analysis if of course used during this stage, but in certain ways that it is worthwhile stressing. Firstly, analysis is based on a "nest" study system in which the various levels of organisation fit inside each other in a pattern fairly similar to that proposed by R.D. Hart. Although farms and the farming systems that they represent form a major study level in this procedure
in that it is at this level that farmers consider most of their decisions, it has also appeared indispensable, particularly in much research carried out in African countries, to pay close attention to territorial and ethnic units (villages, local regions) to which the farms belong and whose social and technical organisation determines a large proportion of the decisions made by farmers. This accounts for the importance awarded to the idea of agrarian system in numerous studies carried out in these countries by French teams.

The other important aspect of this phase is the importance given to analysis of the diversity of methods of exploiting the environment, whether on a regional scale, on that of farm management or that of crops and livestock. Beyond the stratification of the zone investigated or of the resulting farm typology, this analysis of diversity is emerging as a very effective way of revealing differences in behaviour and strategy of farmers and hence the constraints that they encounter in the management of their farming system. The analysis of livestock farming systems in a mountain zone carried out by researchers from SAD-INRA, Toulouse, is a particularly good demonstration of this.

Finally, the work carried out by CIRAD and ORSTOM research teams both in France in the Segala zone, in Egypt and in the Ivory Coast, has shown that historical analysis of agrarian systems and farming systems is a good way of identifying blockages in these systems and of determining the most appropriate ways and means of intervening to ensure their development.

2.2 Experimentation

The second phase of the Research and Development approach aims at experimenting the innovations that are most likely to provide a solution to the problems identified during the initial diagnosis. One of the specific features of the approach is that this experimentation concerns both technical improvements to farming systems and to the organisation of farmers necessary for them to adopt these innovations. The work carried out to date on technical experimentation has made it possible to obtain a fairly broad agreement on the methods to be used. It has concerned in particular the multidisciplinary nature of such research and the need to combine various types of measures both in the research station and in the farming environment with greater or lesser control of the factors involved in the treatments and a varying degree of participation on the part of farmers.

However, improving the efficiency and rigour of experimentation in a farming environment involves paying special attention to some of the following aspects:

- Choice of site is of major importance for the validity and subsequent extension of the results. It is still too frequently handled in function of opportunities rather than in a reasoned manner. Prior analysis of the diversity of agricultural situations and manner of operation of farms is essential base data for a reasoned choice; the weakness of many experimental set-ups is caused by the inadequacy of this prior analysis.

- Taking into account the real conditions under which farmers put into practice, the technical improvements proposed by research is also a point which must be gone into thoroughly both in the design of trial set-ups in the farming environment and in the steering of thematic station research. A study carried out in Burkina Faso on the very low efficiency of improved varieties of sorghum and millet tested in a farming environment shows that feedback from field to station is still inadequate. The establishment of a suitable framework of reference is essential in order to go beyond the analysis stage and to propose effective solutions for the improvement of production. Much can be learned in this respect from the work carried out by IRAT in Brazil.

Finally, the conditions under which experimentation in a farming environment takes place (heterogeneous environments, the need to limit repetitions, etc.) make it difficult to use the methods of interpretation and validation that are generally applied in research stations.

A certain amount of work carried out by French teams has helped to rephrase the problems of the validation of experimentation carried out in farming environments. Two other types of validation are proposed for use with statistical interpretation, which is frequently the only technique used, although it is not always suitable, in particular for analyzing yield variability of farmers' fields:

- agronomic analysis of yield, combining survey and experimentation as has been done for example in the Camargue by LESCA for rice-growing;

- evaluation by farmers of the tests and trials making it possible to take into account criteria other than yield: earliness, taste, etc.

The second type of experimentation undertaken in research and development operations concerns the organisational innovations necessary for farmers to be able to use the technical innovations proposed to them. This type of intervention appears to be particularly necessary when the improvement of farming systems requires regeneration of the natural resources (soil, vegetation, etc.) and better collective organisation of their management, as has been done for the use of rangelands in Mali. This experimentation also makes it possible to find solutions to the constraints which are frequently observed upstream and downstream of production, whether at the stage of supply of inputs or as regards marketing production. The experiments on fertilizer banks, grain banks and collective propagation of seed undertaken at Maradi in Niger are a good demonstration of this point.

Finally, it is in this spirit of overall change of farming systems that R&D operations have been undertaken with the aim of monitoring and progressively transforming farms which will then be used as reference for the extension of the results obtained. This method of research and development, which has been set up in France by the Reseau National d'Experimentation et de Demonstration (RNED) (National Experimentation and Development Network), is in the process of being extrapolated by DSA/CIRAD for use in a number of Latin American countries for developing livestock farming systems.

Be that as it may, the application of a research and development programme which includes organisational innovations requires a number of preliminary conditions:

- the existence of a multidisciplinary team including specialists in agronomy and sociology;
- a political will for organisation of producers;
- trusting cooperation with development institutions;
- social conditions which permit the innovations.

2.3 Extension of the Results of Research and Development

Research and development operations are limited to particular sites: they will only have a significant impact in terms of development if their results are extended and extrapolated to larger areas, particularly on a regional scale. Two questions then arise: the choice of sites for research and development operations (this point has already been mentioned) and also the question of the products to be circulated and the methods of circulation and appropriation of these products. The products to be circulated are of several kinds:

- methods of analysis and diagnosis of the environment;
- technical results obtained from agronomic tests;
- advisory information for technical or economic management of farms;

- forms and methods of organisation of producers with regard both to the mobilisation of the means of production and relations with the economic background.

The mechanisms of extension of the results of research and development are decisive for development; it is therefore necessary to survey them as a subject of research and to perfect them in liaison with farmers and development institutions. It is not possible here to make a detailed analysis of this part of research and development; several essential aspects can nonetheless be stressed.

- The first aspect is that of the adoption by farmers of the appropriate technical and social innovations. It refers to questions of information and hence of communication with farmers, but also to questions concerning the setting up of systems of supply, credit, marketing, etc. It is stressed however that the mechanisms of adopting innovations are not the same for a "simple", "adaptive" innovation, i.e. a one-off, reversible change which does not lead to large-scale modification of the farming system, and for a "complex" innovation which has repercussions throughout the farming system.

- The second aspect of the extension of results of research and development concerns farmers' mastery of the appropriate technical and social innovations. This mastery is necessary to ensure that the innovation is adopted on a lasting basis and refers to the training of farmers. This takes two main forms: technical training and training in management. It may concern all farmers (distributed in homogeneous categories) or village "specialists" responsible for example for the maintenance and allocation of farm machinery.

- Training in the extension of the results of R&D also concerns development officers. It should enable them to acquire skills necessary for greater effectiveness in their operations.

- A particularly advantageous way of making research and development profitable and far-reaching would appear to be the enrichment and defining of extension programmes whose messages are often unsuitable and too normative.

3. CONCLUSION

Although the research and development process raised many hopes, it must be admitted that the results of its application in the field have often not come up to expectations. One of these difficulties of application is of a very general type: it is the institutional keying-in of research and development operations. By definition, the latter overlaps fields of activity and skills of research institutions and development institutions and the linking with one rather than with the other is always a source of tension. The ideal is to set up concertation and cooperation structures between research and development, but latent opposition or the lack of effective cooperation between these institutions sometimes makes these coordination structures difficult. Nevertheless, the respective responsibilities of research and of development are very closely interlocked in the management of an R&D programme.

It is considered that solving this type of problem requires:

- strengthening of institutional links between research, development and training on the lines of the existing situation in certain developing countries, since in many developing countries these complementary functions report to different ministries;

- greater responsibility of farmers in the organisation and management of development and of research, that is to say by the development of professional bodies. In France, these organisations have played a very active role in guiding research for the profit of rural development.

Finally, we should like to finish with a reminder that the research and development procedure is based on the awareness that rural development cannot proceed without the support of farmers. Farmers are the decisive factor in development. Without their support, any attempt at intervention is doomed to failure sooner or later.

A fairly radical change in the present methods and manners of intervention in rural environments is necessary for this awareness to materialise, but full recognition of the farmer as a partner in any research and development operation. This means first of all that his practices and hence his motivation be understood. He must then be involved in the search for new ways of exploiting his environment, and he must finally be awarded an increasing share of responsibility in the organisation of production and its upstream (supplies) and downstream (marketing) requirements. Research and development aims precisely at this.

ATTEMPTS TO INCORPORATE/INSTITUTIONALIZE AN FSR/E APPROACH INTO THE RESEARCH FUNCTION OF AN AGRICULTURAL UNIVERSITY (ViSCA) IN THE PHILIPPINES

by

Ly Tung[1]

1. THE INSTITUTIONAL SETTING

ViSCA is one of the four multi-commodity research centres in the Philippine national research and development network. It is a university-based research centre located in central Philippines (Eastern Visayas Region).

As a multi-commodity research centre, ViSCA has national responsibility for two commodities: abaca and root crops. In addition, it has regional responsibility for 14 commodities, such as coconut, corn, rice, vegetables, beef and chevon, pork and poultry.

With its varied responsibilities, ViSCA conducts a wide spectrum of research activities. As a national centre, its researchers conduct basic, strategic, and applied research on two commodities for which it has national responsibility. As regional centre, it addresses the needs of the Visayas, its service area, for both applied and adaptive research.

The organizational set-up, with the various units involved in research at VISCA, is shown in Figure 1. There are two national research centres: the Philippine Root Crops Research and Training Centre (PRCRTC) and the National Abaca Research Centre (NARC). There are three regional research centres: the Regional Coconut Research Centre (RCRC), the Centre for Social Research (CSR), and the Farm and Resource Management Institute (FARMI) which is the youngest unit. These centres are under the supervision of the Director of Research and Extension.

---

[1] Farm and Resource Management Institute, Visayas State College of Agriculture, Visca

In 1989, there were 201 researchers in ViSCA distributed among PRCRTC (32), CSR (20), FARMI (6), RCRC (6), NARC (5), and the ViSCA technical departments (132). The majority (79%) of the 201 researchers in ViSCA have advanced degrees: 28% with Ph.D. and 51% with M.S. degrees (Gapasin et al, 1990).

2. THE CREATION OF FARMI IN ViSCA-INSTITUTIONAL IMPACT OF FSDP-EV

Attempt to incorporate farming systems research/extension - FSR/E - in the research function of ViSCA has its history with the implementation of the Farming Systems Development Project- Eastern Visayas (FSDP-EV). It is, their fore, in order to have a brief review of the project.

Figure 1: Organizational set-up of VISCA showing various research units

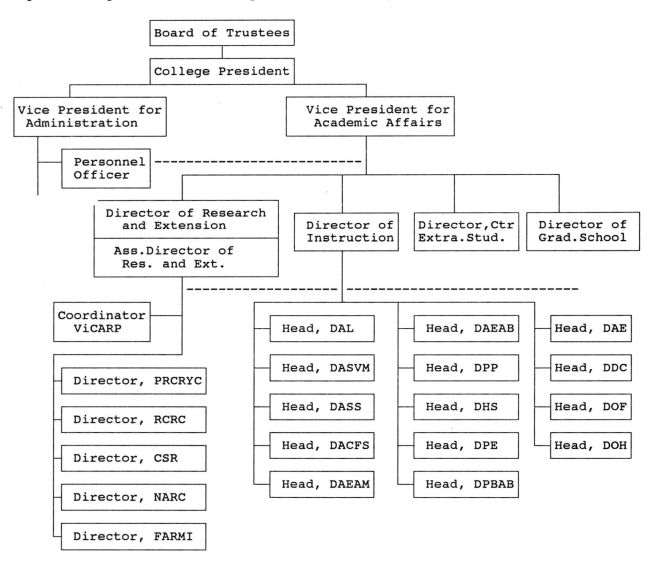

FSDP-EV was initiated late in 1981 with the Regional Department of Agriculture (RDA-Region VIII) taking the lead and ViSCA providing the technical assistance. The project was financed by USAID and the Government of the Philippines.

The project's primary objective was "to establish a mechanism for disseminating rainfed agricultural technologies adapted to the resource conditions found in Region VIII." To accomplish this, the project designers provided for the establishment of six Site Research Management Units (SRMUs) in diverse upland areas of the region. Research and extension activities at each site were carried out by a multi-disciplinary team of the RDA. Also, the project was designed to facilitate the collaboration/linkage between ViSCA and RDA in both research and extension activities for upland farmers. This was accomplished primarily through a scheme by which SRMUs were provided with technical assistance from an ad-hoc group of ViSA faculty members drawn from various technical departments.

The project completed its cycle I in December 1987. USAID, RDA, and ViSCA agreed to continue the project for three more years (1988-1990 or cycle II). This 3-year extension was intended to provide an opportunity for RDA and ViSCA to determine what had been learned through the project and what mechanisms needed to be institutionalized.

In 1988, the RDA decided to adopt the FSR/E approach with the establishment of regular R and E teams at the village level. Such a move by RDA was matched by ViSCA with the creation of a separate unit called the Farm and Resource Management Institute (FARMI). Most of the members of the technical group (ad-hoc) became seconded staff of FARMI. In addition, six plantilla positions were obtained and occupied by six full-time technical staff. The Institute has its own offices and regular budget (a share from the College budget). FARMI is mandated to:

1. Develop/identify, adapt, and disseminate farming technologies to sustain the increased agricultural productivity of the resource-poor upland farmers following the FSR/E philosophy, concepts, and principles.

2. Strengthen and institutionalize the FSR/E approach to research and extension in ViSCA and within the region.

Perhaps at this point in time ViSCA is one of the few educational/research institutions that have created an administratively separate unit to pursue FSR/E. Meanwhile, warning against the establishment of a separate organizational structure for FSR/E is found in recent scientific literature (Birgegard and Fones-Sundell, 1987). The reason for the warning is that, as some authors argue, existing research units will easily see such separate new structure as competitor in terms of resources and functions as well. The next section deals with various issues related to the creation of FARMI and the incorporation of FSR/E in ViSCA.

3. THE INSTITUTIONALIZATION OF FSR/E IN ViSCA - AN ATTEMPT

On December 15, 1989, FARMI initiated a seminar-workshop in ViSCA entitled "Institutionalization of FSR/E in ViSCA." The rationale for this activity is as follows (Ly Tung et al, 1989):

> The research system in ViSCA in terms of organization and management was already complex before FARMI was created in January 1987. One of the strong justifications for FARMI's creation is based on lessons/experiences gained from the FSDP-EV project. FARMI is thought of as an "institutional innovation" to promote in ViSCA the FSR/E. The approach is believed to be more responsive in meeting the technological requirements of the small, limited-resource farmers and in enhancing linkage between research, technology transfer, and farmer.
>
> While FARMI has been quite effective as a promoter of the FSR/E in the past three years (1987-89) with substantial project funds, it is not certain as to what mechanism needs to be strengthened/instituted to sustain the FSR/E approach in ViSCA beyond the project.
>
> To find an effective mechanism is no small task since the ViSCA's research system becomes more complex with the addition of FARMI. It is within this premise that the VISCA department/centre heads and selected senior staff are invited to this seminar-workshop.

The workshop proceedings were fully documented (Ly Tung et al, 1989). Several good issues were brought to the fore during the seminar-workshop; some of the more critical ones relative to institutionalization were as follows:

a. Definition of FSR/E - Only one or multiple definitions?

Now, what is FSR/E in ViSCA today (1989)? Since the early years of the 80's, where are we now in terms of institutionalizing the approach? What are the problems that we continue to encounter? What are the problems that we have successfully solved? And where do we go from here? In trying to look at the solutions to these questions I raised, allow me to raise some issues. One, there seems to be a feeling that there should be only one definition of FSR/E. Is this appropriate? Up to today, as I have attended several meetings in Farming Systems, people continue to debate on what is Farming Systems after a decade of implementation. Why? Do we really need to have only one definition? Or is it appropriate to have multiple definitions. What is desirable, one or many? E.R. Ponce, Director of Research and Extension

b. **What to institutionalize? - The philosophies, the set of values and beliefs, not the label (FSR/E) and the steps.**

The next issue I raise is: What are the central themes or philosophical constructs that we should institutionalize? Sometimes we are too preoccupied with labels, with steps rather than the philosophies, the set of values and beliefs that we should institutionalize. Having identified these if ever we will agree (I hope that we can have certain agreements), how do we institutionalize these in the total R & D program of the college or within each unit involved? And what should then be the role of FARMI? E.R. Ponce.

c. **How to develop proper values and attitudes in our staff?**

But when we talk about institutionalizing we should bear in mind that it is also a process of developing a set of beliefs and values founded on certain theme for our personnel involved in R & D. We have a number of seminars, workshops, etc. regarding FSR/E and what it should be. Is this enough to develop the proper values and attitudes in our staff? Then the second point on institutionalization is: Where are we in the development of proper mechanisms and structures and the allocation of necessary resources which will enable our R & D personnel to collaborate among themselves... E.R. Ponce.

d. **Aside from FARMI, some other units in ViSCA also practise FSR/E. So FARMI should never think that FSR/E FSR/E. Institutionalization in ViSCA is synonymous with FARMI.**

Also, aside from FARMI, we have academic departments and research centres participating in FSR/E. Some are brought about through FARMI's efforts, some do not involve FARMI. So FARMI must never think that FSR/E institutionalization in ViSCA is synonymous with FARMI. There are developments side by side with the activities of FARMI that go on and we have to piece this puzzle together to have a holistic look at the dynamic development of FSR/E, those initiated by FARMI and those activities that happen exclusive of FARMI. E.R. Ponce

e. "The decision to adopt FSR/E as a paradigm for research in ViSCA should not be made through administrative fiat but should be based solely on FSR/E's merits." E.B. Saz, representing all research centres.

f. "We need to be convinced by FARMI that they have the evidence to prove that, indeed, FSR/E is superior; therefore, worth adopting by the rest." E.B. Saz.

g. **The integration of the farming systems perspective does not necessitate the creation of a separate and permanent structure that deals solely with FSR/E.**

Finally, the integration of the farming systems perspective into the R and E programs of ViSCA must take as a prerequisite the orientation and persuasion of the research community that FSR/E is the more relevant approach to research and extension. It does not necessitate the creation of a separate and permanent structure that deals solely with FSR/E. Rather, the integration can be productively woven into the existing units, departments, and centres where expertise is based. FARMI, then, has yet to find its niche in ViSCA as it loses its monopoly of FSR/E. E.B. Saz. E.B. Saz

h. **Duplication of the functions of other units by FARMI.**

What seems to be happening now is that both ESR (experiment-station research) and OFR are done by FARMI, the latter in collaboration with RDA. FARMI necessarily has become a broker between the researcher and farmer on research within the FSR agenda. FARMI may want to continue in this role as FSR is adopted college-wide. Assuming that this happens, FARMI competes with ODREx (Office of Director of Research and Extension) in this role. By logic, the latter is the more appropriate unit to do the brokering. In fact, ODREx has a network of linkages far more extensive than that of FARMI. To add to the complication, some departments and centres also have existing linkages with farmers such that they may not, in fact, do not need the brokering of both ODREx and FARMI.

On the ESR side, FARMI duplicates the functions of other units. And assuming FARMI goes into something other than FSR/E, what should it concentrate on and how does it complement with the other centres to avoid duplication of thrusts? E.B. Saz.

As can be seen, for the first time we were able to elicit the varied views of members of the scientific community regarding FSR/E in ViSCA. These views suggest the following:

a. Some other academic units in ViSCA believe that they also practise FSR/E which does not involve FARMI at all. In other words, the FSR/E was already institutionalized even before FARMI attempted to initiate the effort. (To me this is good.)

b. Since we have a problem on FSR/E definition, there are several variations of R/E activities that are emerging in ViSCA, and the implementors/proponents of these activities can always claim that they are doing FSR/E. (To me this is good. Let various FSR/E variations/approaches emerge.)

c. Under ViSCA setting, if we want to institutionalize FSR/E, please do not talk about the label (e.g. FSR/E), the steps (e.g. diagnosis, design, testing, and extension) because the label may change (in fact it is already changing). Instead, let's try to incorporate the philosophies, the set of values and beliefs. (For instance, it may be better to talk about doing research with a Farming Systems Perspective (FSP) and its On-Farm Research (OFR) Component as proposed by Stoop (1987).)

d. FARMI must show "evidence to prove that, indeed, FSR/E is superior; therefore, worth adopting by the rest." (This issue is a very complicated one since FSR/E is designed to complement the conventional way of doing research, not to replace it. In fact, according to Norman and Modiakotla (1990), "FSR/E facilitates a process and does not produce a product by itself. Therein lies the problem and the difficulty of accountability and credibility for FSR/E.")

e. The contention that "the creation in ViSCA of a separate and permanent structure for FSR/E is not necessary," as expressed by members of the ViSCA scientific community, conforms with the views of Birgegard and Fones-Sundell (1987) mentioned earlier. (But since FARMI was already created in January 1987 through a ViSCA Board Resolution, in my view, it is difficult for the ViSCA administration to make changes. It is more practical, however, to find out how FARMI would best fit into the whole research system of ViSCA. The next section deals with this issue.)

4. THE ON-FARM CLIENT-ORIENTED RESEARCH OF FARMI

In view of the economic difficulty of the country, in search for more effective and efficient ways of doing research, and as we are entering the last decade of the 20th century, ViSCA has got to streamline its research program for the next 10 years (1991-2000). A series of workshops to this effect took place in the early part of 1991 involving most members of the ViSCA research community. Here again, the question of how FARMI would fit into the whole research system of ViSCA was dealt with at length.

Finally, an agreement was reached and that is: FARMI shall focus its research thrust on upland agro-ecosystems that have upland rice and corn as base-crops. Like any other commodity research centre, FARMI may (and should) involve the appropriate expertise found in the technical departments in implementing the research thrust assigned to the Institute. Other long-established commodity/discipline research centres in ViSCA such as PRCRTC, NARC, RCRC, and CSR have their own commodity/discipline research thrusts including research activities that may fall under the umbrella of FSR/E.

In carrying out the new thrust given to FARMI, we try to avoid the notion that FARMI is doing FSR/E as a separate science. Instead, we always keep in mind two important key elements in our research effort. One is the Farming Systems Perspective (FSP) which signifies "an approach" and "scientist's attitude" towards agricultural research (Stoop 1987). The other is the On-Farm Research (OFR) component which has the following major objectives:

a. To identify niches for component technologies;
b. To test and adapt technologies under the specific agro-ecological, socioeconomic, and farm management conditions of identified client groups of farmers;
c. To feedback farm-level information to experiment-station research.

How does FARMI operationalize the two key elements mentioned above? The answers are as follows:

a. FARMI staff (both regular and seconded) do not have any problem in understanding the FSP which signifies a change in approach and in scientist's attitude towards agricultural research. This is attributed to the "institutional" impact of the FSDP-EV project which provided staff with the various training opportunities (locally and abroad) and which exposed staff to real field situations/experiences.

b. With regard to the second element, the OFR, FARMI is very keen in looking at itself and finding out how it performs with respect to the three critical aspects of institutionalizing OFR as cited by Merrill-Sands et al (1989), as follows:
    a. integrating on-farm and experiment-station research so that they form a coherent effort;
    b. involving resource-poor farmers actively as clients in the research process;
    c. developing an interdisciplinary systems approach to agricultural research.

These three critical aspects call for an appropriate organization and management of OFR which have received less attention or sometimes are underestimated.

FARMI started implementing an OFR project since January 1990 which yielded various educational experiences. Below are some examples:

a. **Integrating on-farm and experiment-station research**

Our set-up takes care of this aspect very well. FARMI has only six (6) regular technical staff who specialize in OFR. This is (and should be) a very small group. Because of its small size, FARMI needs seconded staff drawn from various technical departments. These seconded staff do both on-farm and experiment-station research. Both regular and seconded staff form into core teams to deal with specific agro- ecosystems (that are upland rice-based and corn-based). Moreover, if the core team identifies researchable areas that cannot be handled by any member of the team, the research topic is channelled to an appropriate researcher in the proper technical department. In effect, we have a set-up composed of a mix of three types of specialization e.g. the OFR-specialized staff, the OFR- and ESR-specialized staff, and the ESR-specialized staff.

b. Involving resource-poor farmers actively as clients in the research process

Here, we will not run out of examples indicating active farmers' involvement. A classic example is their involvement in the diagnostic stage. Another example is their participation in testing and evaluating several cultivars of corn, upland rice, and sweet potato. Another example: Information is elicited from farmers on desirable characteristics of corn, sweet potato, and upland rice cultivars that they prefer. Such information is inputed into the ViSCA's breeding programs. Another example: We learned from farmers how important and critical the weeding practices are in upland rice. This information is being used by researchers in their attempts to introduce interventions for improved weeding practices. Another example: We learned from farmers about their indigenous technique for soil conservation in the uplands (Ly Tung and Alcober, 1991). And subsequently, the testing and adoption by farmers of Vetiver grass technology as an improvement over their indigenous technology came about (Ly Tung and Balina, 1991). Another ~ example: The women group was asked to identify problems relative to farming and to choose potential interventions for testing by themselves. All these examples and other experiences are reported in several issues (Nos. 1 to 8) of an informal Newsletter entitled: "OFR Notes" published by FARMI. Moreover, the Newsletter is designed to perform the feedback role (or advisory role) of OFR which is the most difficult role to achieve in several institutions according to Merrill-Sands and McAllister (1988).

However, we are still struggling with the problems in selecting farmer collaborators who are representative of identified client groups. In one of the activities entitled "Integrating women's concern in upland rice farming systems," we came up with definite criteria for selecting collaborators based on information from a survey. These criteria are (Alcober, 1991):
- one must be engaged in upland rice-based farming systems;
- one must be engaged in tying abaca (Manila hemp) strands as an off-farm income activity ; and
- one must be raising livestock, especially swine.

In other farmer-collaborative activities, most of the time we simply accept volunteers at meetings. To improve on this, we started trying the "wealth ranking" as a research tool designed to refine our procedure for selecting farmer collaborators.

c. Developing an interdisciplinary systems approach to agricultural research

On the aspect of multidisciplinary versus interdisciplinary research, we employ both strategies. For component technology development, the team as a whole plans the research agenda but the execution and evaluation phases are usually left to the concerned disciplines. For adapting/ testing an integrated production package with farmer- collaborators, the whole team is involved in planning, executing, and evaluating the potential interventions.

On the aspect of sustaining the involvement of social scientists in the team, we do not have any difficulty here. However, the contribution of our social scientists in the team can still be strengthened for them to truly assume the lead role in shaping up the research agenda which is often influenced greatly by technical team members. In other words, experienced social scientists are very crucial; otherwise, technical members will still dominate.

5. SUMMARY

a. ViSCA is a multi-commodity research centre which is a university-based centre located in central Philippines (Eastern Visayas Region). As such, it has two national research centres and three regional research centres. In 1989, ViSCA had 201 researchers with advanced degrees (M.S. and Ph.D.) distributed into five research centres (69) and 14 technical departments (132).

b. From 1981 to 1987 ViSCA was involved in the implementation of a regionally-focused FSRE-type project of which the Regional Department of Agriculture was the lead implementing agency.

c. Capitalizing on lessons learned and experiences gained from this project, ViSCA decided to incorporate/institutionalize the FSR/E approach into its research function by creating in January 1987 a separate unit called the Farm and Resource Management Institute (FARMI). FARMI is thought of as an "institutional innovation" to promote in ViSCA the FSR/E approach to agricultural research.

d. In December of 1989, FARMI organized a seminar-workshop in ViSCA entitled "Institutionalization of FSR/E in ViSCA" which department/centre heads and selected senior staff attended. Results of this workshop reveal the following views as expressed by the ViSCA research community: (1) the credibility of FSR/E as promoted by FARMI is questionable, (2) the philosophy, set of values and beliefs must be institutionalized, not the label (e.g. FSR/E), and (3) in order to institutionalize the farming systems perspective, the creation of an administratively separate unit like FARMI is not necessary because it competes for resources and it overlaps with functions of other academic units.

e. In the early part of 1991, a series of workshop to streamline the research program of ViSCA was conducted, and an agreement was reached for the research thrust of FARMI and that is: FARMI shall focus its research thrust on upland agro-ecosystems that have upland rice and corn as the base-crops.

f. In carrying out the new thrust, FARMI tries to avoid the notion that the institute is doing FSR/E as a separate science. Instead, FARMI keeps in mind two important key elements in its research work: (1) the farming systems perspective, and (2) the on-farm research component.

g. The difficulties that FARMI experienced in its attempt to strengthen/institutionalize FSR/E in ViSCA serve as necessary stimuli for the institute to discharge its thrust the best way it can. In May this year (1991), ViSCA was selected as one of the key sites of an upland rice research consortium (Asia) funded by ADB and IRRI. And FARMI was identified to take the lead role in implementing the consortium's research program which necessarily involves several technical departments of ViSCA. This is a manifestation that FARMI is slowly (but surely) gaining recognition based on the merits of its research performance adopting the OFR/FSP approach.

h. Lastly, if FARMI is to remain effective it will be necessary to maintain the presence of strong leadership from scientists with disciplinary breadth and management experience, especially experience in seeking external funding to support FARMI research programs. The core group of technical staff (both regular and seconded) must be strong in their disciplines, broad in interdisciplinary perspective, and good in coordination, to maintain the respect of other university faculty. This group of regular technical staff, however, should be kept small to discourage FARMI from becoming independent.

REFERENCES

Alcober, D.L. 1991. Integrating women's concerns in upland rice FSR. An interim report submitted to IRRI on July 10, 1991.

Birgegard, L.E. and M. Fones-Sundell. 1987. Farming Systems Research - Issues and Problems of Implementation. Issue paper No. 4. Swedish University of Agricultural Science.

Gapasin, D.P., O. Velarde, and K. Stave. 1990. Researcher's profile and activities at the Visayas State College of Agriculture (Philippines). Staff Notes, No. 90-92. ISNAR.

Ly Tung, F.T. Balina, and D. L. Alcober. 1989. Seminar on the institutionalization of FSR/E in ViSCA: a proceedings report. FARMI, ViSCA. Unpublished. 83 pp.

Ly Tung and L. Alcober. 1991. Natural grass strips are preferred.ILEIA Newsletter, Vol. 7, Nos. 1 and 2:27-28, 1991.

Ly Tung and F.T. Balina. 1991. The introduction of Vetivergrass (Vetiveria zizanioides) to improve an indigenous technology for soil and water conservation. Presented during the Regional R & D Highlights at ViSCA on 15-17 July 1991. Unpublished, 12 pp.

Merrill-Sands, D. and J. McAllister. 1988. Strengthening the integration of OFCOR and ESR in NARS. Management lessons from nine country case studies. OFCOR Comparative Study Paper No. 1. ISNAR.

Merrill-Sands, D., P. Ewell, S. Biggs, and J. McAllister. 1989. Issues in institutionalizing on-farm client-oriented research: a review of experiences from nine national agricultural research systems. Staff Notes, No. 89-57. ISNAR.

Norman, D. and E. Modiakgotla. 1990. Ensuring farmer input into the research process within an institutional setting: The case of semi-arid Botswana. ODI Network paper 16. London, U.K.

Stoop, W.A. 1987. Issues in implementing research with a farming systems perspective in NARS. Working paper No. 6. ISNAR.

V

FARMING SYSTEMS, GOVERNMENTS AND NGOs

Overview and Synthesis

The common view of development is that of a process of economic grow promoted by governments. Development agencies have concentrated upon the improvement of government programmes and much of the development literature has followed the same path. As a consequence, most discussions of the institutionalization of the farming systems approach have concerned government agencies. The gradual acceptance of a more people-focused development concept has caused attention to shift towards consideration of the mechanisms necessary to accelerate the formation of self-reliant farmers' organizations. This process has largely been viewed in the context of a partnership between government organisations (GOs) and farmers organisations. The current chapter reminds us that such a partnership is more complex in reality, since non-governmental organizations (NGOs) have amply demonstrated the need for their inclusion on equal terms with the other two parties.

The paper by Farrington and Bebbington discusses the relative strengths and weaknesses of GOs and NGOs in promoting the farming systems concept, identifies the preconditions that must be met if they are to collaborate successfully and suggests future patterns of cooperation. These discussions are based upon an extremely wide spectrum of experience, largely in Latin America and within rural areas that could be classified as complex, diverse and risk-prone (CDR areas).

The potential of GOs and NGOs to apply farming systems approaches can be assessed in terms of their capacity in the following areas:

- ability to draw upon a range of relevant disciplines,

- understanding the complexity of users' rights and obligations over a range of tenure situations,

- viewing systems development in the context of longer-term and more widely-focused change,

- tendency to focus upon several constraints simultaneously, and

- establishment of self-sustaining institutions.

GOs can offer a wide range of examples of multi-disciplinary approaches, but suffer from the sort of attitudinal problems among personnel that have been discussed in previous chapters. They have the advantage of relatively dependable financial resources, but many of their experiences with have been initiated by donors and led by expatriates.

The more indigenously-based NGOs, on the other hand, often suffer from fluctuating budgetary levels tied to short-term projects. Nonetheless, they tend to be much better staffed with social-scientists willing to work with physical scientists in an interdisciplinary manner. There skills in understanding social organization and the process of capacity creation in local communities, together with mastery of certain techniques of field analysis (eg.RRA) have led to their use by GOs in training government staff.

Although GOs work almost exclusively with farmers on farm-level production problems within a relatively non-conflictive family situation, NGOs are experienced in working with competing or antagonistic groups. They have also often been engaged in joint management of natural resources with farmers' groups and GOs. NGOs are usually more sensitive to sectoral and macro issues than GOs concerned with agriculture. Off-farm and non-farm activities are often included in their programmes, and they tend to think in terms of food systems rather than only solving farm production problems. They are also more willing to look at patterns of constraints, rather than simply compartmentalizing problems.

While GOs and commercial firms tend to avoid CDR areas, because of lower returns to investment, NGOs are usually focused on poverty-alleviation and are, therefore, active in these zones. Not only do they supplement the meagre coverage of GOs in these areas, but they also act to stimulate demand for better services and more relevant research.

The contrasting strengths, weaknesses and areas of interest of the two types of organization are complementary in many respects and suggest the potential for closer collaboration. NGO ideas and initiatives, plus their skills in relation to the analysis of farm situations and organization of communal participation, would combine to advantage with governmental budgetary and logistical support plus strength in technical and scientific areas. The breadth of GO influence could allow the scaling-up of successful, but limited, NGO approaches, while these approaches could be further developed using government research capabilities.

Radical changes in the role of the state in the developmental process are currently being precipitated in many developing countries, by acute budgetary and debt-repayment problems. These changes, which are particularly acute in many Latin American and African countries, have been embodied within structural adjustment programmes that will have the consequence of reducing the relative role of GOs in the economy. This situation presents an opportunity to accelerate the institutionalization of the farming systems perspective, since it could catalyze the growth of farmers organizations as an alternative to GOs.

The potential for NGOs to take a major part in these adjustments is obvious. Unfortunately, they seem less aware of the structural change that is taking place than the majority of GOs, who receive the message directly in the form of staffing freezes, falling real wages and budget cuts. The orthodox socialist view of the predominant role of the state still permeates the strategies of many NGOs. A fundamental rethink is necessary if NGOs are to grasp the present opportunity to catalyze the process of institutionalizing the farming systems approach.

INSTITUTIONALIZATION OF FARMING SYSTEMS DEVELOPMENT

Are There Lessons from NGO-Government Links?

by

Farrington J. and A. Bebbington[1]

I. CONCEPTS and PREMISES

I.1 INTRODUCTION

In areas of developing countries characterised by favourable agricultural production conditions and good infrastructure, farming systems tend to have become 'streamlined' down to a small number of major enterprises with only limited interaction among them. The irrigated production of rice and wheat in the Punjab provide examples of this streamlining. Whilst farming systems perspectives can provide important insights in to problems arising in such areas (agrochemical pollution; salinisation), the overall need for such perspectives is more important in areas characterised by combinations of low and unreliable rainfall, poor soils and hilly topography, where an estimated 1 bn persons seek livelihoods, and where interactions within and among crop, livestock and tree components of farming systems are strong.

Government research and extension departments have found it particularly difficult to implement FSR in these complex, diverse and risk-prone (CDR - Chambers et al. (eds.), 1989) areas. Preliminary evidence (Farrington and Biggs, 1990) suggest that non-profit non-governmental organisations (NGOs) of various types seek to establish themselves particularly in these areas, and that income-generation through agriculture-related activities in a resource-conserving fashion is a major focus of their activities.

---

[1] Overseas Development Institute, Regent's College, Inner Circle, Regent's Park, LONDON NW1 4NS

This paper asks whether some of the difficulties faced by government services can be relieved by closer collaboration with NGOs. It does so by reviewing the types of intervention and farmer-support that are needed in order to promote FSR perspectives in CDR areas. It analyses the strengths and weaknesses of government research and extension services (referred to below as GOs) and of NGOs in addressing these, and then identifies the preconditions that have to be met if GOs and NGOs are to collaborate successfully.

I.2 ISSUES IN INSTITUTIONALISATION

According to NORMAN and COLLINSON (1985), two broad issues need to be addressed in attempts to institutionalise FSR:

- 'the development and dissemination of relevant improved technologies and practices

- the implementation of appropriate policy and support systems to create opportunities for improved production systems and to provide conditions for the adoption of technologies already available'.

Progress on these issues depends on a capacity to:

1) draw on a range of relevant disciplines in order to identify and address specific systems constraints or opportunities at farm, community or higher levels, and to articulate these into research agenda.

2) understand the complexities of users' rights and obligations across various land tenure arrangements (privately owned; various modes of rental or sharing; common property resources; open access land) and analyse what bearing these arrangements have on resource flows and complementarities among them.

3) draw out the implications for systems development of wider changes, such as trends in agrarian structure, terms of trade for agricultural products, rural urban migration patterns and development of rural non-farm employment and investment opportunities.

4) remove several constraints simultaneously. These may lie in or across: production technology; input supply; credit; processing and marketing.

5) establish self-sustaining institutions capable of addressing relevant elements of (1)-(4), and of interacting in a supportive and synergistic fashion.

The first three of these are concerned with methods and perspectives, the fourth encompasses action-goals that stem from these perspectives, and the fifth with means of institutionalising these methods and perspectives, we now consider each in turn.

Progress towards institutionalisation[2] is now reviewed against the five requirements outlined in the previous section.

1. Cross-disciplinarity

GOs have introduced a wide range of diagnostic methods integrating the perspectives of different disciplines - often with strong participation from farmers themselves. examples include: the 'sondeo' in Latin American (Hildebrand 1981); joint treks in Nepal (Mathemaand Galt, 1989); rapid research field hearings in Brazil and Indonesia (Knipscheer and Suradisastra, 1986); rapid appraisals of various kinds (see e.g various issues of RRA notes produced by the International Institute for Environment and Development). A growing body of literature indicates how these approaches have succeeded in identifying issues for research that would have been missed by conventional, single-discipline approaches (Norman et al 1988; Worman et al 1990).

Serious questions have arisen, however, over the extent to which approaches of this kind can be institutionalised and scaled up to cover the whole of research institutes' mandated areas. These include:

(i) resistance among directors and senior staff of research institutes to any effort seeking to divert research resources away from the conventional (and 'respectable') discipline or commodity-based organisation of research in which they were trained and have now established their credentials. Examples of such resistance are quoted by Collinson (1988) and have led to a view that natural or social scientists engaged in applied work of this kind are conducting 'inferior' science and should be subject to resource limitations.

(ii) resource constraints (which may or may not be associated with (i), such as per diem rates which do not cover actual reasonable expenditures (Farrington and Mathema 1991), limited access to transport, and limited overall budgets for field work. Many research services are coming to the view that participatory methods are too costly to be implemented over more than a small percentage of the farmers they are mandated to serve -unless, that is, they are able to devise ways of collaborating with other organisations established in

---

[2] This section can do no more than summarise the main trends and indicate the main sources of information that the interested reader may wish to follow up.

rural areas which are capable of this work - such as NGOs. Examples are provided in the second section of this paper.

(iii) the fact that many of the initiatives reported in the literature were taken by expatriate teams, either in isolation or in collaboration with local institutes[3]. This may assist in methodology development, but at the expense of distracting attention from questions of institutionalisation.

NGOs have wide experience in conducting rapid and participatory appraisal, often being commissioned by government departments to provide training in these methods to local staff. In some instances, their focus has differed from that of GO teams in two important respects: their focus has generally been on empowerment, in the sense of enhancing local people's understanding of their situation and of creating local capacities to address problems and opportunities and the skills they have drawn upon have been reacted more toward the understanding of social organisation and less towards technical agricultural issues. However, numerous examples were presented at a recent workshop[4] (Chakraborty et al, 1991; Fernandez, 1991) of ways in which NGOs had been particularly concerned to strengthen GOs through the provision of training in participatory methods to their staff. Box 1 provides an account of the approach adopted by one NGO. Examples are presented mainly from Asia in the first section of this paper, and from S. America in the second of NGO-GO collaboration not only in the implementation of participatory methods in the processes of research, but also in the much wider context of seeking institutional and policy change favourable to the implementation of farming systems perspectives. This second group of approaches is consistent with the institutionalisation of family systems' development (FSD) as it is defined in the background document for this consultation (FAO, 1991).

2. *Perspectives across private land, common property and open-access resources*

GOs have worked almost exclusively with <u>farmers</u> who own land individually or who have individual usufruct rights through tenancy arrangements. The relative conservatism in this respect among GOs concerned with agricultural (ie. mainly crops and livestock) research and extension should be contrasted with the highly innovative approaches to joint management (ie. between rural dwellers, NGOs and State agencies) of natural resources produced in CPRs, and their integration with those produced on private land, pursued by some government agencies responsible

---

[3] Although numerous local initiatives have been reporte, see for instance Abedin and Haque (1980), and Maurya et al (1990).

[4] Asia Regional Workshop on 'NGOs, renewable natural resources management and links with the public sector' held at Hyderabad, India, 16-20 September 1991.

for eg. watershed management and rehabilitation of wastelands (eg. Karnataka State Watershed Development Cell's collaboration with NGOs in watershed management, Bhat and Satish, 1991).

NGOs have been concerned with renewable natural resource management in the wider context, eschewing the notion of a nuclear family unit in which interests are non-conflicting, and paying explicit attention to the rights and obligations of men and women in a variety of social arrangements encompassing private, common property and open access land. Examples include: the work of IUCN in Nepal to promote local level environmental planning (Carew-Reid and Oli, 1991); the work of Auroville in India on land reclamation (Giordano et al, 1991); the various Philippine NGOs (Mag'uugmad Foundation - Cerna and Miclat-Teves, 1991; Mindanao Baptist Rural Life Centre - Watson, 1991) which have developed land management techniques for farmers who do not enjoy secure tenure on sloping land. A particularly novel NGO programme to promote exploitation by the landless of 'common property' areas of water through duck-keeping is detailed in Box 2.

3. *Wider perspectives*

Although there is awareness among some international research centres of the need for FSR to inform, and be informed by, wider changes in macro-economic and agrarian conditions, attempts to identify cases in which **GOs** have taken wider perspectives have yielded few examples (R. Tripp, pers. comm.).

Despite the fact that the majority of **NGOs** operate on a local scale, many have a coherent view of the types of macropolitical, economic and environmental change that are taking place, and the types of agricultural change that are both feasible within this wider context and compatible with their views on longer-term resource-management strategies. In Bangladesh, for instance, The Bangladesh Rural Advancement Committee (BRAC), for instance, has long been aware that agricultural research by GOs has focused on those with secure access to land, to the neglect of the 50% of economically active population in rural areas which is landless. Faced with the near-impossibility of obtaining land for its clients among the rural poor, BRAC has noted that the increasing adoption of high yielding varieties in Bangladesh has increased the returns to other inputs relative to those to land. It has therefore pursued a highly innovative programme incorporating training, credit and social organisation to facilitate access by the landless to 'lumpy' assets such as tubewells, from which they then sell water to established farmers (Box 3).

4. *Simultaneous removal of constraints*

GO perspectives have largely been limited to the identification of crop, livestock or agroforestry <u>production</u> constraints and the subsequent modification of research agenda This is the trend of the numerous examples given in e.g. the CIMMYT E.African farming systems' Newsletter and the (now defunct) University of Florida Farming Systems Support Project Newsletter. A major shortcoming of GOs is the 'departmentalisation' of areas of responsibility: even the best interdisciplinary teams from a research service will have little to say to departments responsible for input supply, processing or marketing and, if they do, the chances of being heard and acted upon are limited.

By contrast, **NGO** are not bound by demarcations of this kind. Within the sphere of production research, cases have been documented in which NGOs rapidly incorporated new lines of investigation to meet emergent needs. In India, for instance, the Bharatiya Agro-Industries Foundation, which pioneered the introduction of frozen semen technology for cross-bred dairy cattle in the 1970s and, to date, has produced approximately 10% of the cross-bred herd, realised at an early stage that the full genetic potential of the cross-breds would not be realised unless improvements in animal health and nutrition were introduced. It therefore embarked on its own programme of R & D on vaccine production to counter the high price and unreliable supply of imported vaccines. BAIF also began research on a number of high protein feed supplements, including tree products and processed crop residues (Satish and Farrington, 1990).

Other cases have been documented in which NGOs, taking more of a food systems than merely a farming systems perspective, have simultaneously removed constraints across crop production, processing and marketing (eg. Catholic Relief Services on sesame in the Gambia - Gilbert, 1990). Many seek, the better ones successfully, to identify and relieve constraints simultaneously across a spectrum of activities (CRS/Gambia on sesame oil production and marketing; Silveira House on credit, fertiliser and HYVs-MacGarry, 1991).[5] Box 4 details the approach taken by the Mennonite Central Committee to soya production in Bangladesh.

5. *Establishment of self-sustaining institutions*

Historically, in areas where agricultural conditions are favourable and infrastructure well-developed, the benefit:cost ratio of designing and introducing new agricultural technology has been higher than in CDR areas. There is some evidence that this process is running into diminishing marginal returns as the incremental impact of new technologies becomes smaller and problems (pollution; salinisation) caused by previous technologies become greater. Nonetheless, the broad pattern is still that diverse farming systems in CDR areas require a higher value of research input per farm and per unit area in order to achieve a given increment in productivity than do favourably-endowed areas. In other words, research is more costly in CDR areas, and the outcomes less predictable. Inevitably, therefore, a disproportionate

---

[5] Note the criticism that NGOs are spread too thinly across a wide range of thematic areas and inadequately skilled to address some technical issues.

amount of the resources available from government's fiscal base are allocated to favourably-endowed areas. Likewise, private commercial organisations concerned with the development and marketing of new technology focus on these rather than on CDR areas. Given the combination of private sector and government research guided to some extent by 'demand-pull' from the more organised and articulate farmers, it is clear that patterns of institutionalisation in favourably-endowed areas have progressed further along the path followed by now industrialised countries than have those in CDR areas.

How to meet the higher costs of developing adoptable technologies for CDR areas in an institutionally sustainable fashion remains an area in which substantial research is required. However, a few pointers already suggest themselves:

1) it is clear that enough resources from government alone are unlikely to be made available to address adequately the range of renewable natural resource management issues requiring attention in CDR areas.

2) in ecologically fragile areas, particularly those crossing national boundaries, or likely to have cross-border or even global effects if poorly managed, a macroeconomic case can be made for long-term support from the international community to the development and implementation of sustainable RNR management practices. The 'debt-for-nature' swap arrangements such as that currently being implemented in the Philippines (Ganapin, 1991) provide an example.

3) under most circumstances, however, a substantial contribution to the identification and solution of RNR management problems in CDR areas -much higher than in favourably-endowed areas - will have to be made by the people living in those areas. Through processes of consultation with rural people and their representative organisations, tasks need to be allocated in ways allowing the most cost-effective use to be made of the skills and facilities that each can offer.

There are scarcely any examples in which GOs have perceived the need for and successfully elicited the local participation essential to (3) above[6]. For most NGOs, however, community participation is essential, from the broadest level of stimulating awareness of its situation, through the screening of options for change, down to the testing and disseminating of particular types of intervention.

Some GOs have succeeded in eliciting local participation in the identification of needs, the initial screening of technology options, the conduct of trials and the evaluation and dissemination of results. These aspects of participation also feature in NGOs' work, but in many cases the processes run deeper: numerous NGOs are

---

[6] A small number of notable exceptions from South America are discussed in the next section.

fundamentally concerned with stimulating the emergence of local membership organisations and helping to create awareness among these groups of the constraints they face (in agriculture, but also much more widely) and the possibilities of overcoming these.

A number of NGOs have been concerned to develop with and among local people the skills and capabilities to address their needs. Numerous examples are available, for instance, of the training given by NGOs to local groups of farmers in the selection, multiplication and distribution of seeds. Examples include: the Catholic Relief services in the Gambia (Gilbert, 1990); the Mennonite Central Committee in Bangladesh with vegetable seeds (Buckland and Graham, 1990); Ramakrishna Mission in India, also with vegetables (Chakraborty et al, (1991)); MASIPAG in the Philippines (Miclat-Teves, forthcoming). Seeds for use in agroforestry systems are produced by a number of NGOs including, for instance, the Nepal Agroforestry Foundation (Pandit, 1981). See also the review of NGOs' agriculture related activities in Nepal by Shrestra (1991).

A complementary approach taken by some NGOs has been to stimulate demand-pull by farmer groups on a range of government services, including research and extension. There can be little doubt that the responsiveness of research and extension services to farmers' needs faces multiple constraints (some of which are reviewed in Box 5), the removal of which is a process requiring long-term effort: farmers' capabilities need to be strengthened on several fronts (identification of own needs; ability to articulate these to GOs; development of sufficient political influence to be taken seriously by GOs). Simultaneously, GOs need to be persuaded to become more open and responsive to the needs of small farmers (a course which may in many respects make their own work more difficult), to set up regular channels through which small farmers' requirements can be expressed, and to strive towards allocations of research resources that are equitable across the various sectors of the farming community - and this balance may well have to be defended against claims on resources vociferously articulated by wealthier farmers.

The stimulation of demand-pull by farmers on research agenda has a strongly intuitive logic as part of the process of making government institutions more responsive to farmers' needs, and replicates processes that have evolved over a century or more in the North. In this sense, it may also be seen as part of a wider set of processes towards democratisation. However, the substantial changes and developments that are needed on both sides as part of this process imply a lengthy evolution towards this goal. This accounts, first, for the limited number of cases in which research institutes and farmers have engaged in dialogue in this context, and, second, for the fact that NGOs are likely to have to represent small farmers' interests in 'pulling' services from GOs for a lengthy transitional period - examples of which are cited below.

Whilst it would be premature to draw firm conclusions about whether, and how soon, 'bottom-up' pressures towards greater responsiveness by research and extension services to farmers' needs will occur, there are already very clear indications that the most progressive NGOs, government organisations and farmers' groups are engaged - albeit in many cases in the early stages - in processes leading towards this goal. It is not, therefore, premature to suggest that the types of management, responsiveness and accountability required of GOs if they are to respond to these emergent demands will be very different from those found hitherto. Nor is it premature to analyse the processes by which current practices might be modified towards those required for the future if FSR & D is to be institutionalised on a more self-sustaining basis. It is towards these questions that we turn in the second section of this paper.

BOX 1

NGOs and Food System Perspectives: MCC and Soya in Bangladesh

The Mennonite Central Committee is the international service agency of the Mennonite and Brethren in Christ Churches of the USA. It currently has over 1000 N. American volunteers working in over 50 countries. Operational since 1973, MCC's Bangladesh agricultural programme comprises 18 MCC volunteers working with 135 Bangladeshi staff with a budget in 1990 of US $400,000 (excluding expatriate maintenance costs).

The MCC Agriculture Programme has been operational for 17 years. Compatibility with the Government's Five Year Plans is a fundamental basis for planning, although MCC's activities have a particular emphasis on the rural poor. In its initial stages the Agriculture Programme was seen as an attempt to create sustainable links between the available government research capacity on the one hand and the extension apparatus and the cultivator on the other. As the programme has diversified, some have focused more on government collaborations than others.

MCC researchers have had long-standing relationships with the Bangladesh Rice Research Institute (BRRI), the Bangladesh Agricultural Research Institute (BARI) and the Bangladesh Agricultural Development Corporation (BADC).

Over 500 ha of soyabeans are now produced annually in Bangladesh, which is almost entirely the result of MCC's efforts. Involvement with GoB institutions on soya stems from 1975, when the government setup the Bangladesh Coordinated Soybean Research Project (BCSRP), made up of seven institutions, a food corporation, and MCC. This was coordinated by the bangladesh Agriculture Research Council (BARC). GoB involvement in soybean research was spearheaded by BAU where varietal trials were conducted as well as the development of inoculum for soybean that could be domestically produced. BADC was also involved in seed multiplication, and BCSIR in soyfood development.

BCSRP was terminated in 1981, the two main factors being: difficulty in producing good (soybean) seed, and the need for solvent extraction facilities. These production problems aside, INTSOY's (International Soybean Programme of the University of Illinois) closure of production promotion must also be considered a factor. After the termination of the project, only BAU and MCC continued soybean research to any significant extent.

The termination of BCSRP in 1981 put most of the responsibility for the establishment of soybeans onto MCC in conjunction with BAU. With clear evidence that the establishment process would require an integrated approach including both supply and demand-side interventions, it obligated MCC to attempt a large-scale effort.

BOX 1 *(Continued)*

Presently MCC's soybean effort is quite large, involving not only agronomic varietal research, seed multiplication and extension, but also market promotion, and soy-food product development: an integrated approach to crop promotion. This approach has evolved over time from primary emphasis on the supply-side factors, combined with homestead level demand promotion through cooking demonstrations. At least two breakthroughs can be seen as critical to bringing the soybean effort where it is today. These include the introduction of an Indian soybean variety, Pb-1 in 1985 which has vastly improved seed quality and good seed storability over existing Bangladesh varieties. This variety has led to improved plant stand and allowed farmers to store their own seed.

A second area of recent success associated with the soybean effort has involved the development of demand for soybean by private snack-food companies. MCC increased its effort on the demand-side from 1988 by developing an overall marketing strategy which included approaches to local businesses to encourage them to use soybean in their process. This has led to a situation in the last two years where supply cannot keep pace with demand and farmers are more confident to grow soybeans. In conjunction with rising prices for other pulses, these factors have allowed soybean acreage to expand from 100 ha in 1987, to over 500 ha in 1989.

In the past four years, MCC has been primarily interested (in terms of cooperation with GoB) in having the National Seed board release Pb-1 as an official variety in Bangladesh. This would be a major step to having soybeans established in Bangladesh. However, even with very good agronomic results for Pb-1, MCC could not even attract enough interest in the GoB to have them send a team from the National Seed Board down to see the crop and talk to farmers. This visit would have been a key step in the process of acceptance of a new variety by the Seed Board.

Finally in 1989, soybean was included in the Crop Diversification Programme (CDP), a CIDA-GoB jointly sponsored programme. A five-year soybean action plan was drawn up, and soybeans finally seem to have become a part of the Ministry of Agriculture's overall crop promotion strategy.

Source: Buckland and Graham, 1990

BOX 2

DUCK RAISING and the EXPLOITATION of COMMON PROPERTY WATER RESOURCES in BANGLADESH

Friends in Village Development in Bangladesh (FIVDB) has been involved in the promotion of duck rearing and development and adaptation of related technologies in Bangladesh since the early 1980s. Ducks can be reared in the areas of flat, wet tropical terrain found throughout Bangladesh, with the exception of the Chittagong Hill Tracts and the barind areas of North Bengal.

FIVDB has pioneered duck extension work in Bangladesh by importing foreign breeds and has incorporated duck rearing into its income generating strategy for poor rural women. In rural Sylhet there are now thousands of duck farmers raising improved varieties of ducks such as Khaki Campbell, Cherry Valley 2000 and the recently introduced Ding Zeng, in flocks ranging from 10 to 500 birds. A recent FIVDB survey indicated that about 350,000 improved ducks are now being raised.

FIVDB has worked to develop and introduce associated technologies into its project area, through establishing contacts with agencies working on ducks in other parts of the world. For example, FIVDB has developed an appropriate hatchery system based on a Chinese model.

There have been several areas of collaboration between FIVDB and public sector research institutions eg. research into duck disease at the Bangladesh Agricultural University. FIVDB has also given training to groups of duck raisers in other areas of the country, and improved rearing methods are known to have been taken up beyond Sylhet.

FIVDB plans to expand its activities by:

- i) collaborating with a research institute to further enhance and stabilise the improved blood line
- ii) supporting the formation of local groups to promote improved hatcheries on the Chinese model
- iii) expanding the capacity of FIVDB's central hatchery from 30,000 to 100,000 ducklings/yr.
- iv) introducing earthworm production as a new high-protein feed source
- v) collaborating with duck raisers and extension workers to improve current training programmes.

The programme has a highly specific focus in two contexts:

- i) as a 'backyard' activity, it focuses on rural women who customarily do not engage in income-generating activities outside the homestead.
- ii) ducks are taken (often by children) to common property areas of open water each day. The income-generating possibilities they offer are therefore particularly suited to the approximately 50% of the rural population in Bangladesh that is landless.

Source: Ahmed, Z. 1991.

BOX 3

INFLUENCING the QUALITY of GO TRAINING and EXTENSION:

The Case of Aga Khan Rural Support Project (Gujarat, India)

AKRSP was established in 1985 to promote community participation in Natural Resources Management through local groups with the objective of enhancing income levels among the rural poor. A major component of its work has been to enhance the performance of government services at the local level, and to increase local capacity to draw on these services. AKRSP's efforts to enhance training and extension services have been modified in the light of experience since 1985. Broadly, they fall into two periods:

1985-87

In response to requests from local communities, AKRSP arranged for extension and training visits from the State-level agricultural university and from the Directorate of Agriculture. Despite efforts to familiarise GO staff with field conditions, feedback from farmers on the training sessions revealed that they were too theoretical and 'blueprint' in character, remote from local experience and delivered in a top-down 'lecturing' style. Initially AKRSP made little headway in getting GO staff to respond to these comments.

1988 onwards

AKRSP felt that more participatory approaches were needed, but would first have to be developed at grassroots level before being introduced to GO staff. It therefore developed a methodology of Participatory Training and Extension, involving farmers in resource appraisal and technology generation, technology adaptation and testing and technology diffusion. It also worked on developing a cadre of local village extension volunteers to innovate, experiment and conduct extension at the village level and focused on using farmers as trainers to ensure a strong component of farmer to farmer extension in the training and extension programmes.

AKRSP was now confident that these approaches could be disseminated to the state training and extension institutions.

If then started working again with state training institutions and government departments through a series of workshops and consultation meetings with the officials and trainers at all levels. Over a period of more than two years, this process has resulted in building an enabling environment within GOs to try new participatory methods, to develop processes by which training becomes effective and to set up a semi permanent network of trainers which meets to discuss issues of common thematic concern. A consultative committee comprising of State departments, the Agricultural University and other NGOs has been set up to tackle difficulties as they arise and to consider how to increase the effectiveness of training and the extension efforts. This is leading to field testing of new methods and institutionalisation of these methods within the state training and extension institutions. This demonstrates the importance of networking and emphasises that an NGO must gain first hand experience before trying to introduce any process innovation into public sector agencies.

Source: Shah and Mani, 1991

BOX 4

NGO INNOVATION in the CONTEXT of CHANGING AGRARIAN STRUCTURES - Bangladesh Rural Advancement Committee and Tubewell Irrigator Groups

BRAC is a national development agency with multifaceted programmes and activities spread across the country. It has been active since 1972 when relief efforts were begun in the aftermath of the liberation war. Its objectives are poverty alleviation and the empowerment of the very poor, including women. BRAC focuses on people and has a target group approach.

BRAC takes a holistic approach to development with people as the subject and the different sectors as the object: organisation and human development, primary productive sectors, credit and health-care. Interventions are programmed for specific fields to attain the objectives of poverty alleviation and empowerment of the poor.

Different existing technologies (both hardware and knowledge) are being used as tools in economic activity. BRAC does not see a role for itself in advanced research into technology, which is better left to specialised agencies. Instead it identifies in conjunction with its target groups simple and low cost technologies which can be used by the poor for their own benefit and for the benefit of society at large. The poor can own, operate and manage these technologies to generate employment and income, increase productive efficiency of different sectors and to empower themselves.

The use of groundwater for irrigation first started in 1961 with 50 government-managed deep tubewells. For the second 5-Year Plan (1980-85) the area under lift irrigation was envisaged to rise rapidly from approximately 1m ha to almost 3m ha, and a policy of privatisation of pumps was adopted to overcome some of the severe operational difficulties that had been faced by governmental and para-statal agencies.

Two NGOs, Proshika and BRAC, saw considerable scope for the landless groups with whom they had been working to take advantage of the privatisation policy. Whilst both had been campaigning for better access by the landless to land, they began to realise that land values were increasing as HYVs were gradually adopted, so that the prospects of land redistribution were becoming more remote. At the same time, however, HYVs placed additional demands on other inputs, particularly on reliable water supplies. Therefore, as some opportunities were narrowing for the landless, others were opening up.

By providing group credit for pump purchase, and training in its operation and maintenance, BRAC enabled landless groups to acquire 110 deep tubewells, 32 shallow wells, 3 low lift pumps and 1 floating pump in the two years up to 1990, the number of deep tubewells being expected to reach 900 by 1992.

Some problems remain, particularly in the provision of after-sales service and repairs to pumps, and in collaborating with GOs to enforce recommended distances between pumps. Nevertheless, this innovative work by NGOs in making 'lumpy' technology available to the rural poor illustrates both their capacity to identify technical possibilities within the context of changing agrarian structures, and their comparative advantage in organising groups capable of acquiring and operating the technology.

Source: Mustafa et al., 1991

BOX 5

SUMMARY OF CONSTRAINTS LIMITING THE RESPONSIVENESS OF RESEARCH AND EXTENSION INSTITUTIONS TO FARMERS' NEEDS

Four types of constraint can be identified. Several of them have already been treated extensively in the literature and so are given only brief mention here.

1. Excessive centralisation of decision-making authority

   Many of the consequences of the cumbersome and time-consuming procedures prevalent in many research services have been documented in the OFCOR (Merrill-Sands and Kaimowitz 1990) and Research-Extension linkage studies (Kaimowitz (ed) 1990) conducted from ISNAR. Further practical examples of the way in which research stations are not merely prevented by over-centralisation from adopting systems approaches but are ineffectual in practically every sense are provided from Nepal by Farrington and Mathema (1991). The second part of this paper addresses in some detail the options for decentralisation that can be explored jointly with NGOs.

2. Inappropriate reward systems

   Kaimowitz (ed) (1990) has discussed in some depth how researchers and extensionists are discouraged from responding to farmers' needs by reward systems that have the effect of making them responsive largely to the internal bureaucratic hierarchy. Classically inappropriate reward systems include the annual assessment of researchers' performance almost solely on the basis of the volume of published output and widespread in e.g. South Asia. Discussions by one of the authors (J.F.) with a group of middle-level Indian researchers suggested that two principal criteria are widely used by journal editors in that country when considering manuscripts for publication: first, whether the yield levels obtained exceeded the previous highest; second, whether a novel experimental technique was used. It is not difficult to imagine how researchers are being influenced by such criteria into initiating laboratory-based, single-commodity research, with consequent bias against field-based systems-related work.

   Reward systems are not easy to change, but recent moves in the Philippines Ecosystems Research and Development Bureau (Tomboc personal communication 1991) have succeeded in reducing the role of publications-based criteria in staff assessment and increasing the role of criteria based on responsiveness - whether to the needs of farmers or of other scientists.

3. Inflexible programming and budgeting procedures

   Government research institutes in many countries receive funding allocations only for agreed research programmes and projects. To have their entire funding locked into pre-determined programmes severely limits their capacity to respond to the technical problems that arise on a day-to-day basis in NGO agriculture programmes. Participants at the recent Asia Regional Workshop on NGOs, renewable natural resources management and links with the public sector (Hyderabad, India, 16-20 September 1991) felt that, at least in the Indian context, the possibility should be explored of allowing 5% of research institutes' funds to remain 'unallocated' with the intention of facilitating flexible responses by scientists to requests for assistance from the various types of agency involved in implementing natural resources management programmes, including NGOs.

BOX 5 *(Continued)*

4. <u>Inadequate fluidity of staff movements between government staff and NGOs.</u>

   The case studies from S.America reviewed in the second part of this paper indicate how informal contact between NGO and GO staff is a prerequisite to more formal linkages, and how the movement of staff from government to NGO, and vice-versa, has facilitated such contact. This fluidity is in stark contrast with the paucity of inter-agency staff movements in e.g. S.Asia. Major inhibiting factors there appear to be the rigid requirements for entry into government service, and the structure of emoluments in government, a large part of total emoluments being made up by e.g. health, housing and pensions benefits. There are currently few provisions for government staff to leave the service temporarily and then rejoin without losing accumulated benefits of this kind (and 'promotion points'). However, steps have recently been taken to facilitate the temporary secondment of government offices to the private commercial sector, and participants at the Asia Regional Workshop on NGO-GO links mentioned above were concerned that similar provisions should be developed for two-way staff exchanges between GOs and NGOs.

II. INSTITUTIONALIZING FARMING SYSTEMS PERSPECTIVES IN SOUTH AMERICA: Learning from NGO successes and problems

II.1 INTRODUCTION

The foregoing has discussed the areas in which progress must be made in order to foster the institutionalisation of farming systems perspectives in agricultural research and development. We have also suggested that NGOs have made significant contributions and innovations in each of these themes. In order to deepen this analysis, and this suggestion, in this second major section of the paper we assume a more circumscribed regional focus, that of the Andean countries and Chile in South America[7]. The justification of this focus is that these countries offer cases of well developed NGO sectors which have accumulated experiences of a more general relevance. Perhaps of more interest, though, is that in several of these countries, moves have been made by public sector research and extension institutions to develop closer links with NGOs to take advantage of what the state perceives as their comparative advantages, and thus to incorporate NGOs into a national research and extension system with a systems and participatory perspective. Although still in their early stages, these initiatives are of interest for they illuminate both potentials and pitfalls in a co-ordinated strategy.

In this section, then, we analyse the experiences of NGOs in FSR/D and participatory agricultural development, what they indicate about how to operationalise and institutionalise these concepts, and the comparative advantages of NGOs in this. This leads into a discussion of the reasons for inter-institutional contact, and some of the reasons for current interest in it - from the perspectives of both NGO and public sectors. We shall then look at the lessons for institutionalisation that can be learnt from the different contacts thus far initiated, and consider promising types of relationship. The discussion suggests that in the end, the appropriate changes for institutionalisation should not revolve around the question of how to combine public sector and NGOs; and far less should we be thinking of how far it is possible to devolve as many tasks as possible to NGOs. Rather the driving question should be 'what can be learnt from NGO experiences about the characteristics of an institutional structure that meets the demands of an FSR/FSD perspective?' These experiences suggest that appropriate characteristics should prioritise:

---

[7] The material comes from case studies written for an on-going research project in Bolivia, Chile, Colombia, Ecuador and Peru. The project is being co-ordinated jointly by Anthony Bebbington, Penny Davies, Martin Prager and Graham Thiele. As will be clear in the text, Bebbington's discussions with NGO and public sector staff in the region have also been extremely helpful in preparing this paper.

a) the promotion of farmer participation in agenda setting and decision making. This should not only be a consultative participation, but rather a participation in which those farmer organisations play an increasing role as an equally responsible partner in the administration of research and extension systems;

b) mechanisms for increasing local institutional flexibility, allowing:

   (i) adaptation to local needs and technology development initiatives; and

   (ii) work on <u>food systems</u> issues (eg. product processing, marketing, linking technology transfer to local co-operative and credit systems etc) and not only on the <u>farm level</u> dimensions of farming systems;

c) institutional decentralisation, with real financial and decision making power being given to local institutions.

The institutionalisation of farming systems perspectives in South America has much more to do with institutional decentralisation and deconcentration, and rather less to do with the complexities of modelling production systems, or garnering and analysing vast amounts of data within a systems framework. In this regard, and although in many respects it confounds any effort to strengthen agricultural development institutions, the current barrage of policies oriented to reducing the size of the Latin American state could offer real potential for institutionalising FSD. However, this will be a far less statist version of FSD than that assumed to date by most authors and NGOs.

II.2 FARMING SYSTEMS DEVELOPMENT AND NGOs IN LATIN AMERICA

II.2.1 Systems perspectives in NGOs: rhetoric, reality and needs

Most of the strongest NGOs in the study region now use the rhetoric of farming systems. They talk of ecosystems, of interactions between production practices, of complexities, of holism and of integrated perspectives (eg. CCTA, 1991; Field & Chiriboga, 1984; Berdegue & Escobar, 1990). To consider how successfully NGOs have moved beyond the rhetoric, to operationalise farming systems perspectives is to assess their response to several of the challenges of institutionalisation proposed in the preceding section, namely: the adoption of cross disciplinary forms of analysis and diagnosis; the simultaneous removal of constraints; and the adoption of perspectives that go beyond both the level of the private, family owned farm, and beyond the farm gate.

Conceptual diversity in NGOs

First, however, it is important to consider NGOs' conceptualization of 'system' - for here we see great divergence, both between NGOs and within them. In a recent workshop on Andean production systems, one of the conclusions was that the different non-governmental, public and international institutions present were in fact using quite different concepts of system (INIAP-IDRC, 1991). Some meant the crop-environment system, others the environment-society interface; some meant production system alone, others the food system (cf. Rhoades, 1988). Yet the terminology was often the same.

These different concepts can co-exist within an NGO, inspiring both debate and internal critique. In the Chilean NGO, the Grupo de Investigaciones Agrarias (GIA), there have been two competing concepts of farming system (Sotomayor, 1991; Nazif et al, 1990). The one more in the vein of Anglo-American FSR, strongly quantitative, model based and 'measuring everything in sight'; the other based on more descriptive analysis of farming systems, tracing its roots to a French tradition that, in part because of the post-1973 diaspora, has had considerable influence among Chileans who have subsequently returned to work in rural development. Adherents of the latter perspective have criticised the former for spending unnecessary resources collecting and modelling data into systems models that in the end are far too complex to be directly useful for GIA's development work.

The tension between farming system specification and NGO activism: a need for research support

Beyond these conceptual debates, and at times confusions, is a perhaps deeper problem -the lack of data on which to base a thorough systems perspective. Overview studies in the Bolivian altiplano and Peru have identified the lack of such data as a serious constraint (CCTA, 1991; Kohl, 1991; Gonzalez & Miranda, 1991; Bebbington, 1989). Such lack of data can lead organisations that speak of systems, farmers' experiments local knowledge and the like to generate technologies that seem quite inappropriate for local farming systems. In the Bolivian altiplano, for instance, supposedly systems oriented NGOs spent much time and effort generating and disseminating protected crops technologies (greenhouses etc.) that were in several respects inappropriate for both socio-economic reasons (families lacked the time required to instal and manage them, and the technologies were far too costly) and technical-environmental reasons, such as the lack of water (Kohl, 1991). The Chilean NGO Agraria, while working with a concept of regional production systems, admits that in its early years it lacked the data to sustain the concept and so generated inappropriate technologies (Aguirre & Namdar, 1991).

Many NGOs have tried to address this lack of data, although a recurrent difficulty is that donor agencies rarely support such research initiatives. In some cases, these attempts have been successful. In AGRARIA's case, a doctoral student worked with the NGO in the typification of production systems, and the result of this work was to re-orient research foci and to adapt technologies that have subsequently had an important impact in local small farm production (Aguirre & Namdar, 1991).

The problem of gathering the data necessary to sustain a reliable assessment of local farming and food systems is not, however, simply one of scarce financial support for research. It is also an institutional problem - for the long term research necessary implies a dynamic which is at odds with the NGO's concern to respond to rural needs, and develop closer links to the local peasantry. Such links cannot be sustained for very long by research alone - farmers need to see more than questionnaires and research plots if they are to trust an institution.

The contrast between the two dynamics can lead to tensions within the NGO. In the early 1980s, the Ecuadorian NGO, the Centro Andino de Accion Popular (CAAP) began its own experimental farm. The object of the farm was to conduct a form of farming systems research absent in the public sector - namely an evaluation of the different crop rotations and crop varieties that CAAP had identified in its sondeo of local production practices. CAAP eschewed a focus on single commodities taken out of their context: rather they sought to evaluate varieties within these local rotations. On the basis of this evaluation, it was argued, CAAP would be better able to make recommendations based on existing technologies. However, on the one hand this research generated vast amounts of data; on the other hand it required several years before it could give results. Together, these led to two problems in CAAP. Firstly, researchers have had great problems in knowing how to analyse the vast amounts data they have. Secondly, and more significantly, parts of the NGO became increasingly frustrated at the delay before the investment of time, money and space yielded any fruits. This led some to call for the closure of the farm, on the grounds that CAAP should be offering alternatives to the local peasantry far more quickly.

The example is significant because it is not isolated. The Programa Campesino Alternativo de Desarrollo (PROCADE), involving 15 NGOs, and an important inter-institutional technology generation initiative in the Bolivian altiplano, is a second example where NGOs' concern to act has been an obstacle to the operationalisation of a systems research perspective. Although PROCADE's technology generation initiative was explicitly based on a farming systems perspective, an internal evaluation concluded that much remained to be done in order to turn the rhetoric into reality (Miranda, technical director of PROCADE, pers. com, 1991). Despite the concern to address several constraints at once, the evaluation concluded that the majority of participating NGOs were still concentrating on one crop alone, and linking technology adaptation and transfer work with credit lines aimed at the wider diffusion of isolated crop technologies. In the end, the NGOs' primary concern to act impinged on their commitment to a systems method.

However, there <u>are</u> cases of successful long term, data intensive system based approaches to technology generation in NGOs. This is perhaps especially so in the agroecological NGOs, where staff are ideologically united behind the perspective. An influential case here is that of the Centro de Educacion y Tecnologia (CET) in Chile which, after 6 years of rotation research in an experimental agroecological production unit, is now close to making empirically grounded technical recommendations. The risk in such cases, however, where a research focus is sustained by theoretical conviction, is that the internal coherence of the research into production systems is gained at the expense of diminished appropriateness for the actual contexts of peasant production (Carroll, 1992; Kohl, 1991).

The pattern in these examples appears to be the following. While the more academic NGOs are relatively comfortable with the research effort that goes into system specification and analysis, among NGOs with a commitment to acting, it remains difficult to adopt the methods of what we might call 'basic' systems research and technology generation. This is because a relatively pure systems approach implies data intensive research. This in turn requires time and personnel, which are both scarce resources for NGOs committed to intervention, and strongly oriented to building links with the peasantry.

In such cases, action-oriented NGOs require the support of other institutions to assist with the research and data analysis necessary for a refined specification of farming systems. Here universities have a clear research role to play. This research might be done independently of NGOs, the university making results available through usual channels of publication and theses. Conversely, it could be executed in conjunction with NGOs - for example through collaborative research agreements where the NGO assists the university researchers and students with access and entry into the zone, and sensitization to the types of information to look for. In turn, the university could orient the focus of its research more specifically to NGO concerns.

There are two reasons why the second, more direct NGO-university collaboration is preferable. First of all, given the time lag before publication, and the frequent lack of funds for producing documents, a direct contact will speed the availability of results for field actions. Secondly, and more importantly, is the question of finance. In the countries under research, there exist funds for university research for which NGOs are increasingly beginning to compete due to the lack of research funding from their traditional donors. These levels of competition are not necessarily desirable, since they imply the diversion of these funds away from teaching centres, thus reducing the amount of research done in universities, and the contribution of research to the formation of professionals (which also has a key role to play in the institutionalisation of farming systems perspectives - see below). However, if there were such direct NGO-university collaborations, these funds could be used to facilitate farming systems development actions while at the same time retaining their educational roles.

NGO responses to system constraints: the need for technological support

A successful incorporation of systems methodologies does not obviate the need for a commodity specific research capacity. Farming systems perspectives ought lead to a reorientation of the types of research that are done, but they still require experimental stations and specialised researchers.

NGOs will offer little in terms of basic research in any reoriented NARS. Indeed, they themselves need the support of that research, for they generally lack the time, resources and personnel for this type of costly study. They therefore require an appropriate back-up support.

Most NGOs are quite honest and explicit in their recognition of this limitation. The Central Ecuatoriana de Servicios Agricolas (CESA), in Ecuador sees its role as the adaptation and scaling down of public sector technologies for transfer to peasant production systems (CESA, 1980; Mastrocola et al., 1991), leaving to the public sector the task of more basic research. Ideally, CESA sees its subsequent task as being to influence the research agenda in the public sector, in order that its technologies are from the start more oriented to the realities of peasant production systems.

Institutionalising FSR would, then, require a non-NGO sector to continue assuming responsibility for controlled crop research. The issue would be how to instal mechanisms that would require these researchers to orient their work to needs identified by a more systemic and farmer-centred perspective at a field level. To date, it is the failures rather than the successes of NGOs in influencing this station research that hold lessons for future strategies aimed at institutionalising FSR/D.

Overall, NGOs have had limited contact with experiment station researchers. The most frequent contact is the pre-season and one-off meeting to purchase basic, registered or certified seed. Occasionally NGOs have used such contacts to express opinions about how research might be reoriented, but lacking any formal means for exercising any sanction they usually have no influence.

Two experiences suggest possible mechanisms for influencing station research. The one is a donor led relationship, and the example comes once again from CESA in Ecuador. Swiss Technical Cooperation (STC) has since the early/mid 1980s supported a fruit development programme for small producers in CESA and another NGO, the Fondo Ecuatoriano Progressi Populorum (FEPP). At the same time it supported the establishment of a fruit research programme in the public sector NARS, the Instituto Nacional de Investigaciones Agropecuarias (INIAP). Through its position as financier, and its close relationships with CESA,[8] STC was able to foster mechanisms to facilitate CESA's influence over the fruit programme. This

---

[8] At this time STC's Ecuadorian offices were located in CESA's office.

was primarily done via joint research planning meetings. STC also suggested strongly to CESA that it would benefit from the direct support of INIAP, and at the same time CESA's semi-formal representation in INIAP's programme allowed it to make stronger complaints when this support was not forthcoming. Despite ups and downs in the relationship, CESA cites this as a case where it was able to reorient INIAP research toward the economic and agro-ecological constraints of peasant producers.

A second example shows a case of public sector (rather than NGO) initiative in Santa Cruz, Bolivia (Bojanic, 1991; Thiele et al. 1988). Here, two factors contributed to public sector interest in collaboration with NGOs. Budget cutbacks in the mid-1980s weakened state extension services, leaving the public sector research organisation, the Centro de Investigacion Agricola Tropical (CIAT), without an adequate extension service. After 20 years of experimenting with different forms of extension, this only strengthened CIAT's increasing awareness of the chronic difficulty of operating a public sector extension service of 20 to 30 staff in a sparsely populated area of 370,000 square km without the involvement of varying types of local organisation in taking responsibility for many elements of the extension process. The size, number and quality of local organisations (NGOs, farmers' organisation) meant that they eventually came to overshadow the government extension service, especially after budget cutbacks.

At the same time, several staff of CIAT had close personal links with local NGOs, and some had previously worked in these organisations. The director of CIAT was also on the Board of Bolivia's largest NGO, the Centro de Investigacion y Promocion del Campesinado (CIPCA), which was also active in Santa Cruz. CIAT needed the NGOs to do extension, and give field level feedback. Likewise, because they were operating in a colonization zone where knowledge of viable alternatives was still germinal, the NGOs needed research support. What made the collaboration possible was that both sectors had personal reasons to trust the other. Out of this came an arrangement in which NGOs participate in CIAT's annual research planning weeks, and in the planning of the regionalised 'research-extension linkage units' (Thiele et al. 1988). Once again, through a semi-formal representation within the public institution's research planning process, NGOs were able to exercise some influence over CIAT so that its work be more oriented to peasant production systems.

Nonetheless, these examples are few, and even in these cases NGOs feel the research institutions only responded partially to peasant production problems. This still limited success suggests that institutional arrangements are required in which the NGO and other local organisations have more formal power to influence agenda setting in research institutions. This implies voting power and direct participation. While this may seem like one of the solutions that 'would almost certainly be infeasible, both for practical and political reasons' (FAO, 1991), there are reasons to suggest why this may not be so. These reasons, along with possible mechanisms will be discussed in section C.

Wider system perspectives addressing multiple constraints: an NGO forte

Regardless of the problems mentioned above, there are many senses in which NGOs show great sensitivity to systems concerns, if not the rigors of systems research methods. In this sense their work is based on the premise that, **ab inicio**, several constraints, both on-farm and off, must be addressed simultaneously.

In part this sensitivity reflects the fact that NGOs have achieved a degree of cross-disciplinarity in their teams that public sector on-farm research and development teams rarely have. In Ecuador, for instance, in the late 1970s the public sector INIAP had planned to have a social scientist in each of its on-farm research teams - currently it has none. In Chile, the Pinochet government's Instituto de Investigaciones Agropecuarias (INIA) had little interest in small farmers, and so consequently had no interest in anthropologists sensitive to their needs! By contrast, NGO teams typically combine agronomists, economists, sociologists and anthropologists. Some have more technical scientists, others more social scientists. If anything, some NGOs acknowledge an excessive social orientation in this work, and accept that in the future they need to strengthen their agronomic and veterinary capacities (Jaime Borja, director of CAAP, pers. com. 1990). But all combine disciplines in their teams, and the socio-political positions of their agronomists tend to make them half-caste social scientists.

The effect of this is a far greater tendency to work on several constraints at once. While some donors point out that this has its disadvantages, because it may mean doing a little of many things in a rather superficial way (Avanthay, representative of STC in Ecuador, pers. com., 1991), this is a problem that can be managed with care in selecting a restricted set of areas on which to act (which in turn requires access to the sort of research allowing careful specification of farming systems, as discussed above). Nonetheless, Avanthay's comment is important, and privately many NGOs acknowledge a tendency to involve themselves in too many areas at the same time. In the case of CESA this realization has led to a recent effort to improve their planning techniques, in order to fine tune and better target their technology adaptation and extension activities.

Perhaps more important than simultaneous work with different elements of the production system is the far greater tendency of NGOs (and peasant organisations) to combine on farm and off-farm concerns in their work. Among several Chilean NGOs, for instance, this combination is reflected in their attempts to combine ideas of production systems and what some of them call 'micro-regions' (eg. Sotomayor, 1991). The idea of micro-region here is an attempt to integrate on and off-farm themes, in the context of regional ecological and socio-economic conditions, without reaching the greater issues (and generalizations) raised by the concept of a region[9]. In this regard the idea bears some resemblance to the idea of food system. In all these formulations, what one sees is a desire to combine crop production and other issues, such as product transformation, marketing, and credit.

Combining such concerns has the additional effect that, by increasing on-farm rates of cash accumulation the propensity and ability of farmers to adopt technical improvements is also enhanced - because of increased monetary income and reduced monetary risks.

In Ecuador, the public sector on-farm research programme has no contact with institutions dealing with material and administrative technologies for product transformation - nor does it have contact with credit lines. This can be contrasted with NGOs that combine these activities. In TTP in the province of Cotopaxi, CESA has for several years worked with an Indian federation in the improvement of barley production technologies. Recently, encouraged by the federation itself, CESA has sought outside assistance to help plan the installation of a peasant controlled barley milling plant, to process the greater quantity of higher quality barley now being produced. In Salinas, also in Ecuador, FEPP has worked with the local federation of communities combining cheese production and marketing, with credit and technical assistance to help with the improvement of livestock and pastures - and subsequently, along with other NGOs, it has begun to work with the federation in forest management technologies to respond to overgrazing and deforestation pressures that were the result of the economic success of cheese production and livestock-pasture development.

A final example comes from the Alto Beni of Bolivia, where another peasant federation (El Ceibo), that began with the goal of improving the marketing channels for peasant cocoa producers in the Beni, has subsequently combined these activities with a technology screening, adaptation and extension programme for the generation of organic cocoa production technologies specifically oriented to a particular market - European demands for health foods (Trujillo, 1991). This is a clear case of one form of non-public organisation, joining forces with other NGOs, to develop production technologies in a way that is specifically oriented to off-farm concerns, and which together yield a significantly increased income for the producer.

The importance of this food systems perspective is clear - as is the gravity of the fact that it is largely absent in the public sector. Such a perspective, combining technology generation with concerns for product transformation, has a direct effect on income, thus reducing pressures to out-migration and so increasing local labour availability for resource management and conservation. A recent study of peasant organisations in Ecuador has identified this connection between on and off-farm

---

[9] The smaller micro-region is also a more realistic concept for an institution the size of an NGO, which alone could never address regional issues.

technologies as one that should receive prime attention in the future (Bebbington et al., 1991).

An especially interesting case of this combination of on-farm and off-farm perspectives in an NGO comes from the experience of AGRARIA in Chile. Prior to 1990, AGRARIA had preoccupied itself with technology adaptation and extension, with a particular interest in wheat based production systems. This was a role understood explicitly as meeting a need (technical assistance to the peasantry) left deliberately unattended by the Pinochet regime. With the transition to democracy the new government has assumed responsibility for financing peasant technical assistance and will meet this by contracting other organisations (among which are NGOs) to perform the role (see below). However, the state still has no agency providing specific assistance in post-harvest technologies, or marketing etc.

AGRARIA has taken advantage of this change in government programmes to develop project strategies based on a wider systems perspective that incorporates on and off-farm concerns. First of all, AGRARIA re-evaluated how it will use its own financial resources. The conclusion has been to reduce the resources it dedicates to farm level system development, and to finance this work by winning state contracts for the provision of technical assistance. Instead, AGRARIA has oriented its own resources to off-farm stages of the food system and in particular to post-harvest activities. Chilean tax law is such that if a farmer can sell wheat directly to the parastatal wheat purchasing agency, rather than via a grain trader, s/he can recover a value added tax, as well as gain a higher price. For small peasant farmers acting alone, this is usually beyond their possibilities, because of the small amounts sold, the distances involved etc. More importantly, until recently the agency has only been present in certain regions and not others. Seeking to change this situation that worked to the disadvantage of the small farmer, AGRARIA has intervened in one region to assess the feasibility of its playing a bulking, sales and tax recovery role for the peasantry, and the possibility of the agency extending its presence to additional areas of significant peasant production.

Initial experiences have been favourable, and suggest that the activity might both cover the costs AGRARIA incurs as intermediary while also increasing farmer incomes. In addition, the increased income to the farmer facilitates the adoption of the wheat technologies that AGRARIA has screened and validated, and with which it works in its technical assistance. Consequently, AGRARIA has sought to expand the activity to other areas, using its additional income from the technical assistance contracts to cover start-up costs. However, the step toward institutionalisation goes further than this because, having seen the experience, the Instituto de Desarrollo Agropecuario (INDAP) the state agency responsible for credit and technology transfer, has itself commissioned a study from AGRARIA to assess how far the model could be replicated in other parts of the country. This state interest reflects both the success in AGRARIA's projects and the increasingly social democratic orientation of the state.[10]

The first examples, of CESA, FEPP and El Ceibo, were individual cases of how NGOs have systems perspectives that go beyond the farmgate. While they attract attention they always leave an observer dissatisfied that they remain very local experiences. Indeed, the individual NGO can do little, <u>directly</u>, to widen the model - in Annis' (1987) terms to 'scale it up.' However, the case of AGRARIA suggests that these food systems perspectives could be institutionalised on a wider scale through a reorientation of the way public institutions work, incorporating lessons and models from NGOs' innovations. Significantly, for the public sector to incorporate such localized, or regionalised NGO innovations, it itself requires a regionalised institutional structure. In other words, for the public sector to scale up NGOs' experiences, it will have to be reorganised along the lines of the structure of NGOs.

In earlier years, of the centralized, so-called bureaucratic authoritarian state (O'Donnell, 1977), in which power was concentrated both spatially, socially and institutionally in the central state, this would have been unthinkable. Today, however, as the state itself moves to decentralise, there is greater possibility of it scaling up such innovations. Just as the adoption of production technologies has required the removal of constraints at the farm level, so the adoption of institutional and administrative technologies has required the removal of constraints to adoption located in the structure and political orientation of the state. The experience at AGRARIA suggests that some of those constraints may now also be being removed.

Adapting to local systems: NGO flexibility

An additional element of NGOs' sensitivity to local systems is their greater flexibility in being able to respond to problems as they arise, learn from their own mistakes, and take into account additional dimensions of the local food systems as they become relevant.

This capacity to be flexible owes much to NGO's institutional structure. While it would be incorrect to say that there is no involvement of the NGO's central office in local decisions, it is generally the case that in NGOs this influence is less rigid, and is executed much more rapidly than in the public sector. This allows field workers more opportunity to follow their own initiative, and more discretion over

---

[10] It ought also be noted that an important factor in these institutional decisions to follow AGRARIA's lead is the presence of AGRARIA staff in the post-Pinochet public sector. One of AGRARIA's founders is on leave and working as Director of technical assistance in INDAP; another, who still works at AGRARIA is on the directorate of COTRISA. The parallels with CIAT, where CIAT's director is on CIPCA's directorate, are clear.

the use of resources.

An illustrative comparison comes from Ecuador and is that between the NGO CESA and the public sector's on-farm research programme, the Programa de Investigacion en Produccion (PIPs). In both organisations the relationship between field staff and programme director is good, and there is a strong willingness on the part of the directors to support field workers. However, there are important differences. On the one hand, in CESA, central office senior staff are responsible for only 3 or 4 field projects and therefore can give them more frequent support than can the Director of the PIPs who must work alone in supporting 9 different field teams. And secondly, to approve field staff's spending requests the Director of the PIPs must gain Ministry of Agriculture approval - which often arrives after the need for local expenses has passed. By contrast, in CESA authority can be granted much more quickly. Moreover, if a PIP worker confronts a problem that does not fall within the PIP mandate, of on-farm trials and some technical assistance, he will not be permitted to spend money in response to it; conversely, CESA field staff have more flexibility to respond to different types of local need. Consequently CESA's field staff can respond more rapidly and effectively to local problems than can the PIPs' staff. The National Director of the PIPs is quite aware of this, and often compares these aspects of CESA's work favourably with his own programme, and would change accordingly if he had authority (Cardoso, 1991).

II.2.2   Farmer participation in NGOs: rhetoric, reality, lessons

Many of the strongest NGOs in South America were formed as a response to, or in the context of, repressive right wing and military governments (Breslin, 1991). Frequently the NGOs represented attempts of left of centre groups to find alternative means of doing politics, and of establishing a relationship with the popular movement in a context where rural unions, peasant organisations, and left wing parties had been made illegal, or were being intimidated (Annis, 1987; Annis & Hakkim, 1988; Lehmann, 1990; Loveman, 1991). Many saw their FSD work as attempts to create small spaces for rural democratization (Bebbington, 1991). Given this socio-political background, it is of little surprise that more than anything else NGOs have stressed the idea and goal of peasant participation in the development process.

Their concept of participation, however, goes beyond the ways in which it is generally treated in farming systems and participatory research literature. Theirs is not the participation in a group trek, or a joint experiment, or on-farm trials (eg. ILEIA, 1989; Chambers, 1989); it is instead the participation of the peasant organisation, the participation of the engaged actor reflecting critically on the conditions of her or his poverty, and on the basis of that reflection influencing the design of strategies to attack this poverty. For these NGOs, to strengthen participation means to work in strengthening peasant organisations and in popular education. The ideas of Freire (1973) have been influential.

In this regard, the emphasis has been on looking for project methodologies and actions that contribute to strengthening the co-ordination between individual producers, and subsequently between communities. In such a context, seed and input distribution systems, irrigation development and management, and the installation of on-farm trials on land that is collectively owned, or at least collectively rented and worked, have become priority areas of action. By creating spaces where joint action is necessary, the hope is to foster the formalization of a peasant organisation (CESA, 1980; Mastrocola et al., 1991).

Similarly, participation is not only about garnering farmers' technological ideas. Instead the concern is that farmers, and subsequently their organisations, participate in the design, monitoring and adaptive management of the whole project. Ultimately the aim is that a self-sustaining organisation be established and that it then continue to manage the project after the NGO begins to pull out (CESA, 1980; CIPCA, 1991). The intention is that from then on the organisation should look to NGOs and other sources for specific, shorter term forms of assistance as and when it is required - that they effectively treat them as consultants (Bebbington et al., 1991).

This is the rhetoric of most NGOs in these countries. It has received considerable criticism over recent years, not least from peasant organisations themselves, who argue that in fact NGOs do not release control of projects sufficiently quickly to the peasantry, and that often they impose their own criteria on project management (CSUTCB, 1990; Kohl, 1991; Bebbington et al., 1991). There is a truth in these criticisms, but they must not be taken too far. NGOs do not necessarily perform as well as their rhetoric would suggest (Loveman, 1991) - but their experience in participatory agricultural development goes well beyond that of the public sector.

One of the most important lessons from these experiences is that participation should occur at several levels at the same time. On the one hand it is a daily participation in the guise of a more egalitarian relationship between field worker and peasant; it is also a periodic participation through formal meetings between field staff and local farmer groups; and finally, and perhaps most importantly, it is institutional participation. By 'institutional participation' is meant the practice of permanent peasant representation in the co-ordinating committee for the particular project. This representation gives voice and vote to the organisation, thus giving it certain power to monitor, direct, criticise and vote against elements of the agricultural project with which it disagrees. By combining participation with accountability in this sense there is greater likelihood that local ideas and needs will be acted upon.

Some NGOs also have formal representation of national peasant organisations on their directorate. Here, however, experiences have been less positive, and some NGOs feel that these positions tend to be filled by 'farmers' who have in fact ceased to farm and have become professional peasant politicians. Lacking a strong field presence their contributions tend to be more ideological and rhetorical, and far less practical and useful[11].

Such forms of participation can make the NGO's job more demanding than it would otherwise be - because the NGO opens itself to the criticism of those with whom it has to work from day to day. Nonetheless, most agree that this generally improves their work. However, there are qualifications. Even left of centre NGOs comment that this participation can become excessively politicised - which has led some NGOs to break their links with particular peasant organisations. It also leads to conflicts when the peasant representation demands actions that, from experience elsewhere, the NGO feels are most likely to be counterproductive.

The main point of conflict, in this regard, concerns the point at which project management should be passed fully to the peasant organisation. NGOs usually wish to keep hold of the project longer than the organisation wishes. NGOs argue (and many peasant leaders acknowledge, often off the record) that local organisations require high levels of professional training and of internal coherence before they should take on the responsibility of project management. Conversely the peasant organisation argues that the NGO is receiving money in the name of the peasantry and then keeps it to itself. It is also worth noting that although NGOs speak of the need for professionalisation and maturity in the peasant organisation, their own excessively 'social' rhetoric has often kept them from giving the organisations the training in business management, accounting, banking and so forth that are necessary for the organisation to take efficient control of the project.

There is truth in each observation. But what is also true is that there are cases where NGOs have worked with peasant organisations, facilitating their professionalisation, and that the organisation has then become a successful, largely self managing enterprise. It is at this point that institutional sustainability, and the institutionalisation of FSR/D has been genuinely achieved - when the farmers organise and finance their own research and extension support systems, act as representative bodies making claims on government as and when appropriate, and continue to press for appropriate national support systems and policies.

There are few examples of this occurring among small producers (there are more among organisations of larger, wealthier producers), but the following is one illustrative case (see Healey, 1988; Trujillo, 1991). In Bolivia, the peasant organisation El Ceibo now has its own technology adaptation, screening and extension service staffed by the peasants themselves. Its workers, who have come from the region, have received professional training in research institutes and universities. Together they now manage an organisation that has generated

---

[11] I refer to interviews in CESA, Ecuador, the Centro de Servicios Agricolas (CESA) in Bolivia, and the National Association of Quinoa Producers (ANAPQUI) in Bolivia.

organic production technologies, has brought in germplasm from other countries, screened and adapted it to local needs and established a farmer-to-farmer extension service. El Ceibo combines these research actions with successful processing, and national and international export operations.

The implications of these experiences for the institutionalisation of FSR/D is that participation must go beyond local group discussions about trial design and evaluation, to include higher level peasant participation in the design and monitoring of FSR/D actions by representatives of the organisation. Combined with professional training for the organisation's own members, this can ultimately lead to the organisation assuming the bulk of the administration and design of its FSR/D project, using the national research and extension institutions, and NGOs, as resources to be consulted, and ultimately contracted, when specific problems arise (Bebbington et al., 1991). In the design and implementation of these mechanisms, NGOs experiences of joint administration will have much to offer.

The possibilities for government research and extension services to work with NGOs in these strategies for institutionalising FSR/D are clear enough. Provided that the government side wishes to become more responsive to small farmers' needs, it can link with NGOs in: (a) furnishing the skills and facilities required to help them in strengthening and professionalising peasant organisations; and (b) allowing peasant organisations (and, at least in the short and mid-term, their NGO mentors) an increasingly influential role in determining government organisations' agenda. In turn the role of the NGO will be to press to ensure that peasant organisations in fact achieve this degree of representation. While some may criticise this intermediary role for NGOs, preferring an immediate and direct representation of peasant organisations, the reality is that public sector institutions will initially find it easier and more acceptable to participate in this way with NGOs rather than with peasant organisations (cf. Lehmann, 1990).

II.2.3. NGOs as Vehicles for the Institutionalisation of FSR perspectives: overview and implications for relations to the public sector

An overview of potential NGO contributions to institutionalisation

This discussion suggests areas in which there is much to learn from NGOs in any effort at institutionalising an FSR perspective in South American agricultural development.

NGOs lessons for local institutional organisation

The main lessons are the following. Although NGOs manage a far from rigorous, somewhat artisanal conception of farming systems research, successful responses to many of the problems of local systems do not depend on rigor so much as on being able to respond quickly to issues that rural families have themselves identified, developing that response on the basis of farmers' own ideas and practices. This adaptiveness hinges on the decentralisation of decision making authority, and on a relative openness to farmers' ideas. The capacity to make this response with relative success also depends on the chance to consult rapidly with people with a range of specializations, either within the same institution or through informal friendships with other local institutions. Finally, local responsiveness is greatly enhanced by divesting the authority to spend certain amounts of money to local offices of the organisation.

This adaptiveness also hinges on a strong commitment and willingness to, in a sense, make one's own work more complicated. The origins of this work mystique are many, but one key factor is the atmosphere of the NGO office. Smallness of size, the sense of doing work that is likely to produce results, the sense of being valued by the central and national office, and the daily discussions about each others' work, all contribute to the consolidation of an institutional sense of commitment to one's work and one's clients. An illustration comes again from the contrast between CESA and the PIPs: the frequent comment of PIP staff, that they feel isolated and often alone in their work, contrasts with the greater (if not always perfect) sense of teamwork and mutual support in the NGO's local offices.

It is also important to note here the question of wages. In Bolivia, Ecuador, and Peru, public sector workers' wages are falling rapidly, and NGO staff earn more. The low wages in the public sector agricultural institutions undermine commitment to the challenge of incorporating systems perspectives - 'why make the job any more demanding if we are paid so little?' Ultimately this can lead staff to spend less time on the research and extension tasks at hand, because they begin spending more of the day in other income generating activities, which might often include share-cropping relationships with farmers. Ultimately they may leave the institution. Indeed staff wastage is a serious problem - INIAP, for instance, now has no PhD's on its staff (Cardoso, pers. com., 1991). This loss of skilled staff, frequently those who have attended FSR courses, frustrates any attempt to institutionalise FSR perspectives.

NGOs as teachers of FSR/D

The accumulated experience in NGOs of participatory FSR could also be an important resource in any future public sector training programme. Perhaps one of the strongest obstacles to any institutionalisation of FSR perspectives in the public sector is the lack of personnel with appropriate training. Moreover, neither the public sector nor, frequently, the universities have sufficient staff, or experience to provide such training. Conversely, NGOs could undertake such training activities, and there are already cases of this. In Chile, for instance, GIA has provided training in systems and participatory perspectives to other NGOs since 1985. Since the transition to democracy it has begun providing similar courses to public sector staff, as have other NGOs such as AGRARIA and CET.

This form of training in short courses is a start, but is only a short term response. In the end, the question of human resources has to be addressed as a longer term problem. This implies that the university courses that produce the agronomists and other technical staff who subsequently go to work in the public sector must also be changed so that they incorporate FSR perspectives, agroecology, participatory rural development etc. Once again NGOs are beginning to play a role in this. On the one hand are a number of bi-lateral relationships in which NGO staff teach on university courses. Of more interest has been the initiative of the Latin American Consortium for Agroecology and Development (CLADES). This consortium groups together one or two NGOs from most Latin American countries which in turn endeavour to form coalitions among other agroecologically inclined NGOs in their country. CLADES has recently signed an agreement with some of the most important agronomy faculties of several Latin American countries. By way of this agreement, CLADES as consortium, through its contact NGOs in each country, and through the donors that support it, has agreed to work with these faculties in developing and implementing agroecological curricula and research agenda. The explicit goal here is to work towards a slow, but profound, change in the culture of the community of Latin American agronomists - for only then, CLADES argue, will many of the perspectives of systems and agroecological approaches be genuinely institutionalised (Altieri & Yurjevic, 1991). Once again, this is a case where NGOs acknowledge that the challenge of institutional change is too great for them if they act alone, but that if they direct their attention towards other institutions that have a greater influence over the direction of institutional change, then they can 'scale up' the impact of their experiences. It is also a reminder that the institutionalisation of new perspectives must be a slow process directed at deeper changes in the culture and social control of agricultural development.

NGOs as partners in FSR/D

A field level collaboration with NGOs could also increase the strength of FSR perspectives in the public sector. NGOs' strong field knowledge, and their proximity to peasant farmers and their organisations, means they could offer a very important form of feedback to station and regional research. They might also be able to help with the installation and administration of technology adaptation trials in certain key locations - such as, for example, the lands of peasant organisations with a broad based membership. Through such collaborations NGOs could also contribute to the incorporation of social scientific perspectives and data into the research planning process.

An overview of NGO limitations and public sector contributions

Despite these potential lessons, it is also quite clear that there are major institutional and organisational shortcomings in the current structure of the NGO 'sector' itself. Unless addressed, these limitations will constrain the role that the sector might play in institutionalising FSR/D. The limitations could be addressed by changes in the institutional organisation of the NGO sector and by particular forms of collaboration with government institutions.

The institutional limitations of the NGO model

The first, and prime reason why this is so must be the incomplete and internally fractured nature of the NGO sector. There are <u>very many</u> NGOs and many of their actions are unco-ordinated (Bebbington, 1991). At worst inter-NGO competition for status and funding exacerbates these conditions - both factors have been noted in the Bolivian altiplano (Kohl, 1991) and in Chile (Loveman, 1991)[12]. At best, poor communication mechanisms mean that NGOs do not share information among themselves. Their frequent focus on local action in order to develop closer links to farmers has the effect of impeding inter-institutional communication. This in turn can foster the duplication of experiments, lost opportunities to multiply successful innovations, and worse still, the replication of mistakes (Bebbington, 1991).

The formation of NGO networks, at regional, national and international scales is one means of addressing these constraints that has become of increasing importance over the last decade. These networks are intended to facilitate information exchange, co-ordination and planning, inter-NGO training, negotiations with larger donors for projects implemented by coalitions of NGOs, and negotiations with the state. There are both general networks (such as UNITAS in Bolivia) and networks specifically oriented to agricultural development concerns. Some focus on co-ordination and information exchange (eg. the Co-ordinating Commission for Andean Technology, CCTA, in Peru, and the Agroforestry Network in Ecuador); others focus on project implementations, such as PROCADE in Bolivia (CCTA, 1991; Gonzalez & Miranda, 1991; CLADES, 1991). Others are organised regionally in the hope of 'pulling' the policy making process to a regional level, and increasing local participation in that process.

Inspiring these initiatives, is the idea is that the network can facilitate negotiations by increasing both the power of the NGOs (through the greater numbers) and the

---

[12] Loveman (1991:15-16) comments: 'Despite some efforts to translate the NGOs' 1980s slogan, 'let's talk to each other,' into meaningful action, empathy among the majority of NGOs as opponents of the military government rarely translated into concrete co-operation or even informal contacts. Personal and organizational jealousies, competition for funds, old political rivalries, and different visions of the political transition to come complicated inter-NGO relations. [In the future] they must pay more attention to relationships among themselves, from information sharing to joint programming and project implementation.'

efficiency of the negotiation by giving the counterpart a single entity with which to negotiate rather than a plethora of different NGOs.

A second limitation to NGOs' capacity to institutionalise FSR/D comes from their tendency to concentrate in 'fashionable' poor areas (such as the department of La Paz in Bolivia, Chimborazo in Ecuador, and the Cauca valley in Colombia) and to be far more sporadically present in other equally poor areas means that they would not be a sufficient vehicle for nationwide institutionalisation (see FAO, 1990 on the inter-provincial concentrations of NGOs in Bolivia). When the National Program for Rural Development in Ecuador (PRONADER) sought NGO partners for the implementation of agricultural and other rural development actions in its 12 new regional projects, in many areas these NGOs were simply not present (Bonifaz, Subsecretary for Rural Development, pers. com.)

A further institutional problem is that, in general, NGOs live on project financing, and many of the weaker NGOs come and go in tune with the availability of funding (Lehmann, 1991). The nature of NGO finance thus militates against their being vehicles of institutionalisation - for this financial dependency itself is a threat to the NGOs own institutionality. Only the strongest NGOs have built up a body of core funds that protect them from these instabilities.

These problems imply that if the lessons NGOs have to teach are to be more broadly disseminated and accumulated, we need a relatively stable national institutional structure that will collect together, disseminate and then incorporate these lessons into a wider institutional form. The public sector offers this vehicle for scaling up successful NGO experiences, with methodologies, institutional structure, technologies etc. Moreover, if they had closer contact with policy making institutions within government, NGOs would be better placed to adapt their technological work to macro-economic and sectoral policies. They would also stand greater chance of influencing programmes and policies.

The technical limitations of the NGO model

Many NGOs face a second set of constraints in doing the type of diagnostic research required to specify farming and food these systems, to operationalise these systems concepts, and to provide the technological research to generate the responses subsequently identified in these farming systems diagnoses.

This suggests that NGOs require the support of the public sector in their own attempts to operationalise FSR at a local level, and in any NGO led attempt to institutionalise FSR on a wider scale. Public sector experiment station research could provide the technical backstopping that NGOs are unable to provide, and on-farm research programmes could provide the types of regionally adapted technical support that NGOs lack time and staff to do. Similarly, universities could be an important source of research support.

II.3 STRUCTURAL ADJUSTMENT IN THE PUBLIC SECTOR AND FSR/D DECENTRALISATION, DECONCENTRATION AND INSTITUTIONAL CHANGE

How feasible are these potential NGO-government collaborations? For these are not auspicious times for the Latin American public sector - and as such, nor are they for hopes of institutionalising FSR. On all sides public sector staff levels and budgets are being cut. In Ecuador, for example, a 1991 National Austerity law froze all public sector spending on wages and research, limiting yet further the work that the national OFPR programme could do (Cardoso, National Director of PIPs, pers. com., 1991).

Whatever one's opinion of such changes, the fact is that they imply a change in our conception of what the public sector is, what it will do, the operational methods it will use, and how it will be financed. These changes represent attempts at 'radical solution[s] involving a major restructuring of the institutions involved' (FAO, 1991). Thus, in disagreement with the background document for this consultation, the institutionalisation of FSR/D is a question of 'radical' changes in the public sector. However, these are changes that themselves emanate from pressures internal to the public sector, and that challenge us to rethink what the term 'radical' has usually meant in such discussions of peasant centred rural development.

In all the study countries, apart from Chile, our sense is that the public sector is beginning to realise this fact more quickly than are NGOs, many of whom still sustain an orthodox socialist conception of the state's role in society. Many of the discussions of how to institutionalise FSR have also operated with such statist conceptions:[13] the supposition appears to have been always that the NARS' public budgets would grow, or at least be constant, in real terms. Current fiscal trends make such a role, and such budgets, impossible. This, then, leaves us with the question of what these new roles, structure and financing mechanisms will be.

In response to the question, NARS are beginning to propose realignments in their relationships with NGOs. These are of interest to the themes of this consultation, but have often been presented in a very state-centred and clumsy way that casts for NGOs a role as mere implementers of public programmes. This will not be the way forward, at least not in the context of current NGO opinions. Nonetheless, the proposals, and other changes in the structure and institutionality of the public sector suggest ways in which contemporary, apparently adverse, trends might be turned into positive changes for the institutionalisation of an FSR perspective.

In this section we first outline some of the proposals that are emerging from the public sectors in several of these countries. These proposals are couched as responses to prior public failures to adopt FSR perspectives. The truth, however,

---

[13] This is not to say that these authors would ever have considered themselves or their analyses socialist.

is that they are just as much, and often more so, responses to structural adjustment programmes oriented toward the 'privatization' of the state (Barsky, 1990; Bebbington, 1991: 21; Breslin, 1991:7). After discussing why NGOs have responded critically to these proposals, we then look at elements in them, and in the context that has given rise to them, that do offer hope as both economically and politically viable contributions to institutionalisation. We also suggest additional strategies that might be feasible ways of strengthening the institutionalisation of FSR in the present conjuncture.

II.3.1. Public sector proposals for links to NGOs and NGO responses

Several of the studies commissioned for the research project forming the basis of this paper draw attention to the state's increased interest in closer collaboration with NGOs in agricultural research and extension (Bojanic, 1991 for Bolivia; Cardoso et al., 1991 for Ecuador; CCTA, 1991 for Peru; and for Chile, Sotomayor, 1991, and Aguirre & Namdar, 1991). The authors writing from an NGO perspective acknowledge that this represents an interesting, and important change with much potential - but they also stress that as they stand, the proposals are unacceptable and must be changed to give NGOs far more influence in what have been to date public sector decisions.

The initiatives in Bolivia, Ecuador and Peru stem from the public sector's acceptance of its own increasing institutional weakness and lack of resources (Bojanic, 1991; Cardoso et al., 1991; CCTA, 1991). Of the three, the Instituto Boliviano de Tecnologia Agropecuaria (IBTA) in Bolivia, is perhaps the weakest of all (ISNAR, 1990). It is also still strongly arranged along commodity lines. In response to these weaknesses, IBTA has proposed a combined attempt to (i) abandon IBTA's previous role in technology transfer in the altiplano and leave this task to NGOs, (ii) introduce system perspectives into its research, and (iii) allow the participation of NGOs and producers organisations in setting research agendas.

Initially, IBTA stressed the first of these three ideas: i.e. the suggestion that NGOs would be responsible for extension, and IBTA would concentrate on research and 'training' of NGO workers in the different packets IBTA was developing. This instrumentalist and superior attitude to NGOs attracted much criticism from those NGOs who saw the initial plan as an effort to use NGOs as a subsidy to public sector operations, without giving them any power. During 1990-91, this continuing criticism appears to have influenced IBTA's attitude, and more effort has been made to give NGOs more authority in setting research agenda. A recent round of meetings in which IBTA sat down with NGOs and producers organisations to set plans for the next few years of research were largely successful, with high levels of NGO participation (but less of peasant organisations) (Bojanic, pers com, 1991). The NGOs also have permanent representation on a small central advisory and monitoring committee in IBTA. Slowly, a more genuine and effective form of NGO participation may be emerging. While a more direct participation of peasant producers' organisations would be preferable, this NGO participation may be an important bridge toward such more direct representation. In the meantime, NGOs are beginning to assume the important role of linking research and field realities, and because of their voting power in agenda setting, are able to exert a demand pull on IBTA.

The Chilean case is somewhat different. The new government inherited a strong, but streamlined NARS (the Instituto de Investigaciones Agropecuarias - INIA) whose work was consciously directed to large farmer needs. The new government entered with the goal of orienting some of the NARS' work toward poorer producers, and its challenge was to reorient this public sector and introduce a small farm systems-based, and on-farm technology generation and transfer programme. Once again it has sought NGO assistance in this, and the aim is to introduce some 55 nuclei in which INIA and NGOs will share responsibilities and decision making authority in the adaptation and transfer of technologies for peasant producers. Each experiment station will have under its responsibility a number of nuclei (called Centres for Technology Adjustment and Transfer - CATTs) that fall into its agroecological zone, and the work done in these will be jointly decided by NGOs, INIA and INDAP. In some regions, NGOs such as GIA are playing an important role in defining the nuclei's boundaries on the basis of their own concepts of farming systems and micro-regions. Extension work will then be funded by INDAP, but contracted to private agencies. Under Pinochet NGOs were not allowed to bid for such contracts, but now they can, and have become some of the largest contractees in the programme[14].

The motivations for these initiatives are several, but one is the state agencies' simple lack of resources. INIAP's on-farm research programme is quite frank about this: if it is to recover the levels of activity it achieved in the early 1980s, then it will need to share resources with NGOs (Cardoso, 1991 pers. com.). It suggests combined on-farm trials, with the PIPs providing assistance in trial design and analysis, and the NGO helping cover some of the research costs, providing time to monitor the plots, local feedback on farmers' ideas and transferring technologies once they have been validated (Cardoso et al., 1991).

Institutional resource constraints such as these are related to the larger issue of controlling the size of the state. Throughout all this the influence of World Bank and other multilateral donor thinking over the different NARS is clear (Cernea, 1988; World Bank, 1991). IBTA's restructuring in Bolivia, INIA's CATTs and INDAP in Chile, and PRONADER in Ecuador are all multilaterally funded, and incorporate the ideas that (i) NGOs are more efficient at implementing agricultural programmes, and (ii) that using them as vehicles for project implementation is desirable because it avoids the creation of more long term financial burdens on the state. While a new public sector programme within a NARS is difficult to close

---

[14] This is particularly so in AGRARIA's case (Aguirre & Namdar, 1991).

once created, funding to NGOs can be easily turned off at the end of a loan or project period.

It would be unfair to suggest that the only reason for interest in NGOs is that stemming from public expenditure concerns. Indeed, the PIPs in Ecuador are interested in working with NGOs because they believe them to have special local knowledge of farmers' concerns, and an ability to communicate with those farmers that are lacking in the public sector. Similarly, among some in INIA there is the belief that NGOs have much to teach the organisation in systems, participatory and other research methods. Nonetheless, the obvious influence of public expenditure cutbacks, combined with the instrumental attitude towards NGOs (that they should help the work of the public sector rather than vice-versa), and the deep-seated culture of NGO opposition to the state (for understandable historical and political reasons) have together put NGOs on their guard as they respond to these government proposals (Bebbington, 1991; Carroll et al., 1991). Equally problematic has been the NGOs' unwillingness to accept that their 1970s based conception of what the state's role in society should be are not adequate for the 1990s (cf. Loveman, 1991; Sotomayor, 1991, makes a similar reflection from within the Chilean NGO sector).

This guardedness and criticism has in fact caused the public sector to modify its posture, as noted in IBTA's case. This change has in turn facilitated closer contacts, for it has given NGOs a greater role in defining the nature of collaborations. This has perhaps been the problem to date - these proposals for more NGO participation in public programmes have, contradictorily, not been devised in a very participatory manner.

Chile is the counter-example here - a case where NGOs were involved in developing the proposals. Consequently NGO response to them has been more open and constructive. NGO staff were members of party-based commissions on agricultural policy that generated proposals in anticipation of the transition to electoral democracy (Loveman, 1991). These proposals have since influenced INIA and INDAP's programmes, and NGO staff have been contracted as consultants in the continuing design of future public sector FSR/D programmes (see Berdegue 1990). Others have moved into the public sector, but sustain communication channels with their former NGOs. Results so far have fallen short of the proposals, and there is some disillusion among NGOs about the limited progress in introducing FSR/D and participatory perspectives in INIA and INDAP. Nonetheless, many continue their contacts with these institutions in two beliefs. Firstly that contact allows greater chance to continue influencing the state;[15] and secondly that the contact itself is a research experience the results of which the NGO can then analyse and disseminate, and through these publications continue contributing to

---

[15] And indeed Chilean NGOs feel that, albeit slowly so, they are succeeding in this, and INDAP is also changing its ideas about how to work with NGOs and farmers organizations.

the identification of changes necessary for the institutionalisation of FSR/D.

Among these Chilean NGOs there is perhaps a greater awareness than in other countries that their future roles will, in a context of combined social democracy and public sector adjustment, differ from the roles they played in the past - although commentators like Loveman (1991) reflect that there remains still much internal re-assessment to do. Henceforth they must eschew their prior localism and isolation, and begin to contribute to national debates, national institution building and the improvement of national policies and programmes. The contact with the public sector that this must imply will have its disadvantages - but, rather than avoid them, these must be accepted, analysed and criticised constructively, from an acknowledgement of the constraints on the actions of the contemporary state.

II.3.2. NGO-public sector linkages: future possibilities for institutionalising FSR

Though these experiences are still young, they point to more and less favourable starting points for NGO-public sector collaboration in the institutionalisation of FSR/D. In what follows we outline some more specific mechanisms and contexts which might contribute to this institutionalisation.

Contracting, consultative, collaborative or collegial modes of participation?[16]

The forms of linkage identified in the research vary from instances where the public sector simply tries to use the NGO to subsidize its work, to ones where it pays the NGO as a contractee to implement state programmes, to relationships where the NGO has increasing power to draw on its own experiences and use these in influencing decisions about the design of national programmes. Several lessons emerge from these different relationships.

Firstly, if the public sector simply tries to use NGOs to implement parts of programmes it has designed unilaterally, NGOs will refuse to participate. They have no need to do so, and will not accept a subordinate relationship to the state.

Secondly, when the state pays NGOs as contractees to implement public programmes, there is more likelihood that they will participate, as has occurred in Chile, and in a sense occurred in the Social Emergency Fund in Bolivia. However, working as a contractee can cause NGOs a number of problems. In the case of AGRARIA in Chile, for example, the stringent conditions of INDAP's contracts to

---

[16] The terminology comes from Stephen Biggs' (1989) four modes of farmer participation in on-farm research. But the terms lend themselves to considering how NGOs might participate in the institutionalisation of FSR.

implement technical assistance programmes have meant that NGO staff are unable to do anything other than the stipulations of the contract. In this case they have had no time for any on-farm research, and in the first year of contracts had to work with individual farmers and were prevented from working with groups, as they had always done in their own work in the aim of promoting peasant organisation. This conflict with prior work practices, and with the practices of those parts of AGRARIA working with independent funding, caused tensions within the organisation. The attraction of the income that could be generated by these contracts caused AGRARIA to spend much time drawing up proposals to win the contracts and then administering the paperwork they implied. The effect in part is that AGRARIA has become more like a simple service agency, and runs the risk of losing its NGO-style qualities[17]. GIA, which also accepted INDAP contracts, has had similar experiences.

Overall, the strictures of a contractual relationship reduce the NGO's ability to innovate -there is simply no time to do so. If, as the research suggests, one of the main roles of NGOs ought be to continue generating innovations for FSR/D and participatory rural development, innovations which the state does not have time to develop, then this is a very serious drawback to such contracted relationships. By and large, the implication is that from an NGO's point of view it is of interest to accept one or two such relationships as a research experience, in order to analyse the nature of the contract relationship itself in order to then make suggestions for its improvement (the strategy that GIA has taken). But they should not be treated as an income generating venture, and it is not desirable to enter many such contracts.

Instead of utilitarian or contracted relationships, more important are collaborations that can genuinely foster institutionalisation of NGO's FSR/D experiences on a wider scale. Such relationships must be ones of consultation between equals, in which decision making power is shared so that NGOs can put force behind their suggestions, and monitor the public sector's response to them. Despite the problems that will inevitably emerge for NGOs, making their work more complicated than it would be if they acted independently, it is crucial that they sustain such contact. Only then can they be informed innovators, and constant critics, making continuing contributions to discussions on institutionalisation. This sharing of power and responsibility demands changes in the attitudes of both NGOs and the state towards each others' role, and towards their own roles.

Fiscal crisis and agricultural research funding - a crisis with possibilities

We have already commented that an institutionalisation of FSR will require a strong, but reoriented research capacity in the experimental stations, and that for

---

[17] This is something the AGRARIA staff have become increasingly worried about over the last few months (Aguirre, 1991 pers. com.).

NGOs or peasant organisations to influence this reorientation they will need formal power in the programming of research agendas. While this power has been all but impossible to acquire in the past, the very financial crises of public research institutions may create new opportunities. As the general tendency towards introducing market mechanisms in public sector services continues, and as NARS themselves seek to separate themselves from full government control and search out new means of raising research funds, one can imagine a greater tendency for them to accept research contracts from private bodies[18].

Such changes can of course have ominous implications (that the wealthiest farmers will essentially buy out public research), but they also create the possibility that NGOs and peasant organisations could contract research from the NARS. This would allow direct control of that particular field of research, but might also allow them to bargain for a voting presence on the central directorate - effectively as shareholders in the NARS. Of course, these organisations would primarily represent their own interests and ideas, but to the considerable extent that these NGOs and peasant organisations have relatively similar demands and perspectives to other NGOs and peasants, they would be a more general voice on the boards of the NARS representing the countries' resource poor farmer. It is only with this sort of presence that the FSR/D concept of 'feedback' stands any chance of becoming institutionally meaningful.

This has implications for how donors fund NGOs and peasant organisations. It implies that they should move toward more core funding rather than project funding (although the current tendency is the opposite), and that part of this core funding should be for contracting agricultural research[19]. In the future, organisations might be increasingly able to finance such research from their own income. For instance, one can imagine a situation in which the organisation moves into off-farm product transformation and marketing activities. The increased income generated would allow the peasant organisation to contract for research out of its own earnings. There are already examples of this, as in the case of the Union of Yuca Producing and Processing Associations (UAPPY) in the province of Manabi in Ecuador, which is contracting cassava research from the local INIAP research station (Poats, pers. com., 1991). In this regard, it is also worth noting that Indian organisations are aware that this union of economic strength and political influence is an area that must receive their attention in the future. As a leader of the national Indian organisation in Ecuador, CONAIE, recently commented: '[i]t's not that what we have done so far has been bad - it has been good. But we

---

[18] For example, in Ecuador INIAP is currently seeking autonomy from the Ministry of Agriculture in order to gain greater control over the acquisition and use of its own financial resources.

[19] Carroll's (1992) study of NGOs came to a similar conclusion, suggesting that donors should provide more support for institutional strengthening in NGOs, through longer term 'partnership' grants.

haven't come up with any responses to economic problems, and I think that now is the moment to sit down and do so - that way [our organisations] will increase their economic power and through that their political power' (Ampam Karakras, quoted in Bebbington et al., forthcoming).

It may also be possible to gain voting power for NGOs and peasant organisations in cases where the NARS wishes to use the NGO to facilitate a national technology development programme (as proposed for Ecuador in Cardoso et al., 1991, and ISNAR, 1989). As we noted, these relationships imply that the NGO will subsidize the NARS. In such a context, the NGO could demand decision making influence over the setting of regional and/or national research agendas in return for its 'subsidy.' In Bolivia, NGOs have demanded precisely this in the restructuring of IBTA, via the representation of the national NGO network, UNITAS. As we noted, they have been somewhat successful in this. Conversely, the NGOs could require that peasant organisations have this power, as CESA has done, in the slightly different context of a combined irrigation and agricultural development programme in the province of Canar, in Ecuador.

Public sector decentralisation: local power, local responsibilities

Another prominent theme in structural adjustment thinking has been that of public sector decentralisation. The idea has received additional support from the new social democratic governments that also, in their rhetoric, argue the case for deconcentrating the public sector (although in Chile it was the Pinochet government that pushed ahead with regionalisation programmes). While many argue that more is said than actually done about decentralisation, the trend is undeniably there[20].

In general terms, decentralisation should imply a restructuring of the public sector, investing increased power at local levels, with greater possibilities for local organisations and interest groups to participate in the decision making process, and to hold public institutions to account. When this is coupled with the presumption of greater financial resources at these local levels (financed in part by reductions in expenditure at central government levels), then such participation becomes yet more significant.

While this decentralisation is a much wider issue than agricultural development, it is of interest to consider its possible implications for the agricultural sector. In this section we will discuss what it implies for on-farm research programmes, and for

---

[20] As examples we may cite the problems in Peruvian regionalisation under the APRA government (1985-1990), the public commitment of the ID government in Ecuador to decentralization that failed to materialize (1988-1992), and the continuing dominance of La Paz in Bolivian public life much to the annoyance of regions such as Santa Cruz.

participatory regional agricultural planning.

On-farm research programmes: a bridge

Public sector on-farm research programmes are in an interesting position between NGOs and the rest of the NARS, and in the relationship between structural adjustment, administrative decentralisation and the institutionalisation of FSR. Their position is precarious and yet it could offer the most important bridge toward an institutionalisation of FSR involving NGOs and the state. To develop these points we shall consider the case of the PIPs in Ecuador.

While public expenditure cutbacks are felt across the board in NARS, the OFPRs are often yet more affected because of their relative institutional weakness. Their younger, more field and peasant oriented staff often have less power within the NARS, and so are less able to defend themselves. Indeed, in the Ecuadorian case, it has been this lack of power that has frustrated the feedback that these programmes were intended to provide to experiment stations (Cardoso et al., 1991; Espinosa et al., 1989): a feedback that should have played a key role in the institutionalisation of FSR in INIAP (Espinosa 1983).

Yet the same qualities that have weakened the authority of PIP staff within INIAP make PIP staff the part of INIAP that is most likely to find it easy to communicate, build alliances and co-ordinate with NGOs (as the director of the PIPs is currently suggesting himself - see Cardoso et al., 1991). For in many respects, there is little difference between the PIPs' perspectives and those of peasant-centred NGOs. Moreover, the PIPs' staff generally have research skills, and the time for research, that are lacking in NGOs. For this reason informal contacts in the management of on-farm trials have occurred between the PIPs and NGOs (see Mastrocola et al., 1991; Cardoso et al., 1991).

Similarly, PIP staff, along with INIAP's Andean crop researchers (who have long felt they are the sole beacons of a systems' perspective among INIAP's crop programmes), have been able to develop close and cordial relationships with peasant organisations, as in the relationships between the Union of Indigenous Communities of Guamote, the PIP-Chimborazo, and the Andean crops programme. The reasons are similar. The stronger concern for these researchers to work closely and primarily with peasant producers, and their increasing ability to communicate as equals with such farmers, have facilitated collaborations in on-farm trials, and in discussions about how INIAP can facilitate the introduction of peasant controlled crop processing facilities in the Union.

Here lie the bases for one means of increasing the power of the PIPs' farming systems and participatory perspective within INIAP. Financial power increases decision making power, but the PIPs will never have financial power within INIAP because, of course, it is INIAP that pays for the PIPs rather than vice-versa. However, if NGOs and peasant organisations were, as suggested in the previous section, to begin financing elements of NARS research and to obtain decision making power, then by combining their pressure with that of the PIPs, there would be more possibility of orienting station research towards farming systems perspectives.

For these reasons, in order to institutionalise FSR, it is important that OFPRs continue to receive support, despite expenditure cutbacks. Decentralisation of public administration may help in this respect: as regional administrations ought be more committed than national offices to their particular region's programme of adaptive and on-farm research, they would be far more likely to sustain it. Equally, if NGOs and peasant organisations gain more decision making power in regional administrations, as we suggest is possible in the next section, then they too might be expected to support resource allocation to the region's OFPR.

Participatory regional agricultural planning: institutionalising the two pillars of FSR/D

If one vehicle for greater institutionalisation of FSR would be to strengthen decentralised technology generation and adaptation activities, another, more distant but stronger vehicle will be the strengthening of mechanisms for participatory regional agricultural planning, within which the agenda setting for regional OFPR would increasingly fall.

While one source of pressure for regionalisation has been the debate about restructuring the role of government, another source has been the NGOs themselves. This is perhaps clearest in Bolivia, where the latter part of the 1980s saw the formation of departmental networks of NGOs. Such networks now exist in the departments of Santa Cruz, Cochabamba, Oruro, Chuquisaca, Beni, and La Paz. One of the specific goals of the network in Santa Cruz, for example, was to develop regional agricultural policy proposals as part of their attempt to increase the weight of in-region decisions in agricultural and rural development plans. In this sense, the formation of departmental NGO networks has tried to 'pull' agricultural planning toward the beneficiaries, and at the same time, through the combined force of the network, strengthen the hand of NGOs in this decision making. In Chile, there is similar talk among NGOs for the formation of regional networks in order to strengthen and coordinate their roles in regional development.

One of the motivations for regionalisation of policy making is that too often central government programmes have been inappropriate for the particular production systems of the regions. Moving decisions to the regions increases the likelihood that programmes will adapt to local complexities; or in other words, that there will be a farming and food systems base to regional programmes - ie. <u>an institutionalisation of one of the two main concerns of FSR/D</u>. In one case, that of INDAP in Chile, this is already beginning. Aside from the interest in incorporating the post-harvest ideas of AGRARIA, in parts of Chile INDAP is also beginning to introduce GIA's concept of micro-regions as the basis for planning local extension and credit activities.

The more that planning and development decisions are made at a local level, the greater the possibility for NGOs and peasant organisations to participate in, and bring their experience to bear on, local development institutions. That is to say, this institutional change offers a chance to increase genuine and influential participation in agricultural development: a participation in planning, in resource allocation, and in the specifics of regional research planning. <u>This, then, is a concern to institutionalise the second main pillar of FSR/D</u>.

Given the growing constraints on public expenditure, this increased strength in agricultural planning would most probably be combined with an increased role for these local organisations in providing financial contributions for certain local development activities (such as research and training). Such contributions would in turn allow yet greater influence over the orientation of local agricultural development institutions.

By the same token, decentralisation of the public sector increases the possibility that institutional lessons from NGOs' experiences will be incorporated into the structuring of the public sector. This might facilitate local participation and institutional flexibility. It might also allow the creation of smaller institutional environments, organised on a more face-to-face footing that, both in casual and more formal settings, allows more cross-disciplinary discussions and knowledge sharing, and more importantly the generation of the sort of work mystique so often encountered in local NGO offices, and so often absent in public sector institutions.

These would be important changes, for they would represent the 'scaling up' of NGO experiences into a wider, regionally oriented, nationally coherent institutional structure. Ultimately, if the NGO experience is ever to have more than patchy, local impacts, it is this institutional change that must occur. And it is at this point that we see that the question is not how to combine NGO and government - rather it is 'what does the NGO experience have to offer for the design of national structures?'

Such institutional changes would not imply an end to the role for NGOs. NGOs would continue as monitors of the changes, as local interest groups holding institutions to account. They would also continue as innovators, generating new ideas for possible scaling up; and they would maintain the role described in the previous section, as financiers of research, with the combined goal of orienting that research to peasant needs, and of strengthening their own ability as local development institutions to continue innovating and responding to their clients' needs.

While these proposals might appear more conjectural and utopian, they are in line with patterns of institutional change that are already visible in South America. On the one hand the public sector is having to rethink its role, reduce its centralization of power and open up to the influence of some key coalitions. On the other hand, and in response to these very changes, NGOs are having to rethink their roles - reducing their revindicative and contestatory positions, changing their conceptions of what the state, and their relationship to it, should be.

The fact of the matter appears to be that the state is slowly incorporating some of the main ideas that NGOs have defended over the last two decades. Now, the challenge to NGOs is (i) to cooperate with the state to look for ways to make those ideas work, and (ii) to look to new ideas and innovations. Otherwise, it will become increasingly difficult to justify their continued existence.

III. CONCLUDING COMMENTS

Farming systems research and development is much more a perspective on agricultural development than it is a fixed methodology, and as such we are here concerned with the institutionalisation of the principles behind this perspective, rather than with precise, often esoteric, methodologies that often have been propounded as FSR/D. As a perspective FSR/D stresses the importance of embracing simultaneously the agronomic and the social, the on-farm and the off, the scientists' knowledge and the farmers'. Above all, it stresses that small farmers should be the beneficiaries of agricultural research and extension, and that therefore they should influence the directions and styles that it takes.

There have been many more attempts to operationalise and institutionalise these principles among NGOs than there have in the public sector. Where they have been successful, these attempts have demonstrated the importance of institutional flexibility and decentralisation. Above all, they teach that farmer participation must occur simultaneously at a number of levels: in the day to day relationship between technicians and farmers, in the periodic organised meetings between local groups and NGO, and very importantly in the direct representation of peasant organisations in institutional decisions about the direction that agricultural research and development projects should take. Ultimately, the concept of 'participation' must include the aim of strengthening peasant organisations so that they take increasing responsibility, both administrative and financial, for the evolution of local agricultural development initiatives.

These successes, however, remain localized, and for this reason alone NGOs can never be the vehicle for a wider institutionalisation of FSR/D. Rather their successes, and their failures, should be seen as lessons for how to organise a wider institutional structure. In this respect, one very important point is that the weaknesses of NGOs relate to the same institutional structure that has yielded their successes. Their localism and relative smallness mean they lack the resources for the research, both agro-socio-economic and technological, necessary for the full operationalisation of a farming systems perspective. Frequently they cannot find from within solutions to the problems they identify and which peasant organisations present to them. Their localism also means that they often duplicate successes and mistakes, and are relatively inefficient at exchanging information among themselves. They therefore require centralizing institutions that can provide the necessary support and that can scale up NGO innovations. Organisations structured along NGO lines therefore require the institutional support that the public sector has traditionally given.

From this NGO experience comes the lesson that institutionalising FSR/D will require both an institutional decentralisation in the public sector modelled on the structure and organisation of NGOs coupled with the types of centralized institutions traditional to the public sector, which could both support, and facilitate the co-ordination between, local offices.

To date, the public sector has shown a dominant tendency to concentration, centralization, and resistance to the sort of opening up to peasant interests and local demands that the NGO experience suggests is vital to any institutionalisation of FSR/D. In certain countries, such as India, this pattern shows little chance of changing. However, current trends in Latin America, where structural adjustment pressures are slowly enforcing a deconcentration and decentralisation of the public sector, suggest that the possibility of incorporating the institutional lessons from NGOs into this wider structure may be growing. Moreover, this decentralisation will facilitate the sorts of peasant participation in agenda setting, planning, and project monitoring that have to date been so elusive in the public sector. Indeed, in several South American countries, government services, however falteringly, are explicitly trying to learn from NGOs, and involve them in wider programmes.

These changes are still germinal, and it remains to be seen what they will achieve. There are many social interests, and inter-institutional jealousies and conflicts that could frustrate the restructuring that we suggest is necessary. Nonetheless, the force of structural adjustment is a heavy one and whatever it ushers in, it will bring radical institutional change of one form or another. While the opportunities are complex, we suggest that there is the potential to use these new spaces for the sort of profound organisational change that really *is* necessary for any institutionalisation of the principles underlying FSR/D that is worth its salt.

REFERENCES

Abedin, Z and Haque, F. 1990. Innovator workshops in Bangladesh, pp.132-135 in Chambers et al (eds) 1990.

Aguirre, F and Namdar, M. 1991. 'Complementaries and tensions in Agraria-State relations in agricultural development: a trajectory'. Paper prepared for 'Taller Regional para America del Sur: Generacion y Transferencia de Tecnologia Agropecuaria; el Papel de las ONGs y el sector Publico'. 2-7th December, 1991. Santa Cruz, Bolivia.

Ahmed, Z. 1991. 'FIVDB/ODI Case Study on Duck Raising'. Paper presented at the Asian Regional Workshop 'NGOs, Natural Resources Management and Linkages with the Public Sector', 16-20 September 1991, Hyderabad, India.

Altieri, M and Yurjevic, A. 1991. 'Influencing north-south and inter-institutional relations in agricultural research and technology transfer in Latin America: the case of CLADES'. Paper prepared for 'Taller Regional para America del Sur: Generacion y Transferencia de Tecnologia Agropecuaria; el Papel de las ONGs y el sector Publico'. 2-7th December, 1991. Santa Cruz, Bolivia.

Annis, S. 1987. 'Spread at the grassroots: Where it comes from and what it means'. *Grassroots Development* (Fall, 1987).

Annis, S and Hakim, P (eds.). 1988. *Direct to the Poor: Grassroots Development in Latin America*. London: Lynne Rienner.

Barksy, O. 1990. *Politicas Agrarias en America Latina*. Santiago: Cedesco.

Bebbington, A. 1991. 'Sharecropping Agricultural Development: the potential for GSO-government collaboration'. *Grassroots Development*, 15(2), pp 20-30.

Bebbington, A. 1989. 'Institutional Options and Multiple Sources of Agricultural Innovation: A case study from Ecuador'. Agricultural Administration (Research and Extension) *Network Paper* 11. London: ODI.

Bebbington, A; Carrasco, H; Peralbo, L; Ramon, G; Torres, V H and Trunillo, J. 1991. *Evaluacion del impacto generado por los proyectos de desarrollo de base auspiciados por la Fundacion Inter-americano en el Ecuador*. Report to Inter-American Foundation, Rossly, USA.

Bebbington, A. el al. 'Strengthening ethnic organizations for economic development: the challenge facing Ecuador's Indian movement'. *Grassroots Development* (forthcoming).

Berdegue, J. 1990. 'NGOs and Farmers' Organisations in Research and Extension in Chile'. Agricultural Administration (Research and Extension) *Network Paper* 19. London: ODI.

Bhat, K.V. and Satish, S. 1991. NGO-GO collaboration in natural resources management in India: a study of Karnataka State Watershed Development Cell's experience. Paper presented at the Asia Regional Workshop 'NGOs, Natural Resources Management and Linkages with the Public Sector', 16-20 September 1991, Hyderabad, India.

Biggs, S. 1989. *Resource-Poor Farmer Participation in Research: A Synthesis of experiences form Nine National Agricultural Research Systems*. OFCOR Comparative Study Paper No 3. The Hague. ISNAR.

Bojanic, A. 1991. 'El Modelo de CIAT de las relaciones sector publico-ONG en la tranferencia de tecnologia y su replicabilidad en Bolivia'. Paper prepared for 'Taller Regional para America del Sur: Generacion y Transferencia de Tecnologia Agropecuaria; el Papel de las ONGs y el sector Publico'. 2-7th December, 1991. Santa Cruz, Bolivia.

Breslin, P. 1991. 'Democracy in the Rest of the Americas'. *Grassroots Development*, 15(2), pp 3-7.

Buckland, Jerry and Graham, Peter. 1990. 'The Mennonite Central Committee's Experience in Agricultural Research and Extension in Bangladesh'. Agricultural Administration (Research and Extension) Network Paper 17. London: ODI.

Cardoso, V H; Caso, C and Vivar, M. 1991. 'A public sector on-farm research programme's informal relationships with NGOs: the PIP's growing interest in collaboration'. Paper prepared for 'Taller Regional para America del Sur: Generacion y Transferencia de Tecnologia Agropecuaria; el Papel de las ONGs y el sector Publico'. 2-7th December, 1991. Santa Cruz, Bolivia.

Carew-Reid, J, and Oli, K P. 1991. Planning by the people: IUCN's approach to local environmental management in Nepal. Paper presented at the Asian Regional Workshop 'NGOs, Natural Resources Management and Linkages with the Public Sector', 16-20 September 1991, Hyderabad, India.

Carroll, T. 1992. *Intermediary NGOs in grassroots Development: Characteristics of Strong Performers*. West Hartford, Ct.: Kumarian Press.

Carroll, T; Humphreys, D and Scurrah, M. 1991. 'Grassroots Support Organizations in Peru'. *Development in Practice*, 1(2).

CCTA, 1991. 'Nongovernmental Organizations and the Public sector in Peru: The experience of the Coordinating Commission for Andean Technology'. Paper prepared for 'Taller Regional para America del Sur: Generacion y Transferencia de Tecnologia Agropecuaria; el Papel de las ONGs y el sector Publico'. 2-7th December, 1991. Santa Cruz, Bolivia.

Cerna, L-L, and Miclat-Teves, A. 1991. Farmers' initiatives in farmer-based extension services and soil and water conservation in the Philippines. Paper presented at the Asia Regional Workshop 'NGOs, Natural Resources Management and Linkages with the Public Sector', 16-20 September 1991, Hyderabad, India.

Cernea, Michael M. 1988. 'Nongovernmental Organizations and Local Development'. World Bank Discussion Paper. Washington, DC: World Bank.

CESA. 1980. *Un Apoyo al Desarrollo Campesino*. Quito: Central Ecuatoriana de Servicios Agricolas.

Chambers R; Pacey, A. and Thrupp, L A. (eds) 1989. Farmer first: Farmer innovation and agricultural research. London: IT Publications.

CIPCA. 1991. *Por Una Bolivia Diferente*. La Paz: CIPCA.

Collinson, M P. 1988. The development of African farming systems: some personal views. Agricultural Administration and Extension 29,1, pp.7-22.

Escobar, G and Berdegue, J. 1990. 'Conceptos y Metodologia para la Tipificacion de Sistemas de Finca: la Experiencia de RIMISP'. pp 13-43 in Escobar, G and Berdegue, J (eds.) *Tipificacion de Sistemas de Produccion Agricola*. Santiago: Red Internacional de Metodologia de Investigacion de Sistemas de Produccion.

Farrington, J and Biggs, S D. 1990. NGOs, agricultural technology and the rural poor. Food Policy 15(6), pp.479-491.

Farrington, J and Mathema, S D. 1991. Managing agricultural research for fragile environments: Amazon and Himalayan case studies. Occasional Paper No.11. Overseas Development Institute, London: ODI.

Fernandez, A. 1991. NGOs and government: a love-hate relationship. Paper presented at the Asia Regional Workshop 'NGOs, Natural Resources Management and Linkages with the Public Sector', 16-20 September 1991, Hyderabad, India.

Field, L and Chiriboga, M. 1984. *Agricultura Andina: Propuesta de Investigacion*. Quito: Centro Andino de Accion Popular.

FAO. 1990. *Inventario de ONGs vinculadas al Desarrollo Agropecuario y Rural*. (by A Peters and P Mendez). La Paz: FAO.

FAO, 1991. 'Expert Consultation on the institutionalization of Farming Systems Development - Background Note'. Manuscript. Rome: FAO.

Ganapin, D. 1991. NGO-GO collaboration in the debt-for-nature swap programme in the Philippines. Paper presented at the Asia Regional Workshop 'NGOs, Natural Resources Management and Linkages with the Public Sector', 16-20 September 1991, Hyderabad, India.

Gilbert, E. 1990. Non-governmental organisations and agricultural research: the experience of the Gambia. Agricultural Administration (Research and Extension) Network Paper No 12, London: ODI.

Giordano, E. 1991. Greenwork at Auroville: from survival to inter-institutional collaboration. Paper presented at the Asia Regional Workshop 'NGOs, Natural Resources Management and Linkages with the Public Sector', 16-20 September 1991, Hyderabad, India.

Gonzalez, W and Miranda, L. 1991. 'PROCADE: un analisis de su rol coordinador de actividades interinstitucionales y de su relacion con el sector publico'. Paper prepared for 'Taller Regional para America del Sur: Generacion y Transferencia de Tecnologia Agropecuaria; el Papel de las ONGs y el sector Publico'. 2-7th December, 1991. Santa Cruz, Bolivia.

Healey, K. 1988. 'From Field to Factory: Vertical Integration in Bolivia'. pp 195-208 in Annis, S and Hakim, P (eds.), 1988. *Direct to the Poor: Grassroots Development in Latin America*. London: Lynne Rienner.

Hildebrand, P E. 1981. Combining disciplines in rapid appraisal: the Sondeo approach. Agricultural Administration, Vol 8, pp.423-432.

INIAP-IDRC. 1991. 'Foro-taller sobre sistemas productivos y acciones de desarrollo en comunidades campesinas alto-andinos de Ecuador'. 21-23 August, 1991. Quito, Ecuador.

Kaimowitz, D. (ed) 1990. Making the link: agricultural research and technology transfer in developing countries. ISNAR/Westview.

Knipscheer, H. and Suradisastra, K. 1986. Farmer participation in Indonesian livestock farming systems by regular research field hearings. Agricultural Administration 22, pp.205-216.

Kohl. 1991. Paper prepared for 'Taller Regional para America del Sur: Generacion y Transferencia de Tecnologia Agropecuaria; el Papel de las ONGs y el sector Publico'. 2-7th December, 1991. Santa Cruz, Bolivia.

Lehmann, A D. 1990. *Democracy and Development in Latin America. Economics, Politics and Religion in the Postwar Period*. Cambridge: Polity Press.

Loveman, B. 1991. 'NGOs and the Transition to Democracy in Chile'. *Grassroots Development*, 15(2), pp 8-19.

MacGarry, B. 1991. Silveira House: propagation of the use of hybrid seed - 1968-83, pp.13-40 in Agricultural Administration (Research and Extension) Network Paper No 24 London: ODI.

Mastrocola, N; Andrade, N and Camacho, M. 1991. 'Multiple mechanisms of NGO-state relations: successes and frustrations in CESA'. Paper prepared for 'Taller Regional para America del Sur: Generacion y Transferencia de Tecnologia Agropecuaria; el Papel de las ONGs y el sector Publico'. 2-7th December, 1991. Santa Cruz, Bolivia.

Mathema, S and Galt, D L. 1989. Appraisal by group trek. pp.68-72 in Chambers et al (eds) 1990.

Maurya, D, Bottrall, D, and Farrington, J. 1990. Improved livelihoods, genetic diversity and farmer participation: a strategy for rice breeding in rainfed areas of India. Experimental Agriculture 24(3), pp.311-320.

Merrill-Sands, D and Kaimowitz, D. 1990. The Technology Triangle: linking farmers, technology transfer agents and agricultural researchers. The Hague: ISNAR.

Miclat-Teves, A. (forthcoming). NGOs and seed production in the Phillipines. Paper commissioned by ODI.

Mustafa, Shams; Rahman, Sanzidur; Sattar, Gulam and Abbasi, Ahmad. 1991. 'Technology Development and Diffusion: A case study of collaboration between BRAC and the Government of Bangladesh'. Paper presented at the Asian Regional Workshop 'NGOs, Natural Resources Management and Linkages with the Public Sector', 16-20 September 1991, Hyderabad, India.

Norman, D., and M. Collinson. (1985). "Farming Systems Research in Theory and Practice", in: Agricultural Systems Research for Developing Countries, Proceedings of an International Workshop, ACIAR Proceedings No 11

Norman, D W; Baker, D; Heinrich G, and Worman, F. 1988. Technology development and farmers' groups: experience from Botswana. Experimantal Agriculture, 24 (3), pp. 321-331.

O'Donnell, G. 1973. *Modernization and Bureaucratic Authoritarianism*. Berkeley: University of California Institute of International Studies.

Pandit, B H. 1991. Case study of the Nepal Agroforestry Foundation. Paper presented at the Asia Regional Workshop 'NGOs, Natural Resources Management and Linkages with the Public Sector', 16-20 September 1991, Hyderabad, India.

Rhoades, R E. 1988. 'Food Systems Research'. in *The Social Sciences at CIP*. Lima: Centro Internacional de la Papa.

Satish, S and Farrington, J. 199 A research-based NGO in India: the Bharatiya Agro-Industries Foundation's cross-bred dairy programme. Agricultural Administration (Research and Extension) Network Paper No 18. London: ODI.

Shah, Parmesh and Mani, P M. 1991. 'Experiences in Participatory Training and Extension Experiences of Aga Khan Rural Support Programme in Gujarat State'. Paper presented at the Asian Regional Workshop 'NGOs, Natural Resources Management and Linkages with the Public Sector', 16-20 September 1991, Hyderabad, India.

Shrestha, N K. 1991. An overview of Non-Governmental Organisations and the agricultural activities iniated by them in Nepal. Paper presented at the Asia Regional Workshop 'NGOs, Natural Resources Management and Linkages with the Public Sector', 16-20 September 1991, Hyderabad, India.

Sotomayor, O. 1991. 'GIA and the new Chilean public sector: the dilemmas of successful NGO influence over the state'. Paper prepared for 'Taller Regional para America del Sur: Generacion y Transferencia de Tecnologia Agropecuaria; el Papel de las ONGs y el sector Publico'. 2-7th December, 1991. Santa Cruz, Bolivia.

Thiele, G; Davies, P and Farrington, J. 1988. 'Strength in Diversity: Innovation in agricultural technology development in Eastern Bolivia'. Agricultural Administration (Research and Extension) *Network Paper* 1. London: ODI.

Trujillo, 1991. 'Extension y investigacion en la Central de Cooperativas El Ceibo Ltda'. Paper prepared for 'Taller Regional para America del Sur: Generacion y Transferencia de Tecnologia Agropecuaria; el Papel de las ONGs y el sector Publico'. 2-7th December, 1991. Santa Cruz, Bolivia.

Watson, H R. 1991. NGO-GO relationships the Rural Life Center's Way. Paper presented at the Asia Regional Workshop 'NGOs, Natural Resources Management and Linkages with the Public Sector', 16-20 September 1991, Hyderabad, India.

World Bank. 1991. *El Banco Mundial y las organizaciones no gubernamentales*. Washington DC: World Bank.

Worman, F, Heinrich, G, Tibone, C and Ntseane, P. 1990. Is farmer input into FSR/E sustainable? The ATIP experience in Botswana. Journal of farming systems Research and Extension 1, 1, pp.17-30.

VI

SOME COUNTRY EXPERIENCES WITH FARMING SYSTEMS APPROACHES

Overview and Synthesis

This chapter contains four papers, three of which relate country experiences with farming systems approaches. The first paper of the chapter summarizes some of the characteristics and problems of the farming systems approaches that have been introduced in the Asia and Pacific Region over the past three decades. In the paper, the author identifies the driving force behind the introduction of farming systems concepts as the poor adoption rates of new technology by farmers. It is evident that this statement applies particularly to the resource-poor farmers in rainfed areas that had not benefitted from the Green Revolution.

The paper draws attention to the introduction into the region of the FSD approach by FAO in 1983. This approach is characterized as stressing the development of poorer farm families, the analysis of the farm and household and the use of interdisciplinary, informal and participative techniques. Despite the fact that almost a decade has passed since the introduction of farming systems approaches into the region, most research still neglects the socio-economic aspects of technology generation and application. This situation calls for improved staffing patterns - either through recruitment or secondments. Extension services are more knowledgeable about farmers' circumstances, but continue to promote single commodity/package of practices advice that ignores enterprise interactions and resource limitations. The situation is not likely to change until better methodologies are developed to deal with the greater complexity of farm-household situations and community-level considerations.

Dar's paper describes the experience in the Philippines, where the systems approach is widely accepted but there is little agreement on exactly which elements constitute the approach. Similarly, the concept of inter-disciplinarity does not appear to be challenged, yet few instances of true inter-disciplinary work can be found. The roots of the problem can be traced to the fact that agricultural development is still centred upon increasing farm productivity through the application of new technology. This ensures a technology bias that leads to the neglect of socio-economic aspects of agricultural development. The usual weak linkages between extension and research services can also be observed in the

Philippines. There is a need to conceptualize more clearly the role of the extensionist and the farmer in the research process and to develop a shared perspective of agricultural and rural development.

The paper describes the experience of 5 different agricultural development projects in the Philippines, and goes on to examine the improved approaches incorporated into the Research and Outreach Project operated by the Department of Agriculture over the past 4 years. Many of the lessons learned in previous projects are reflected. The most radical change concerns a reversal of the order of the research - extension linkage. An extension-led research framework has been developed, one that view non-adoption of extension messages as caused by factors beyond the commonly alleged conservatism and ignorance of the farming community. Researchers and extension workers working in the project, attempt to learn from farmers, so Rapid Rural Appraisal and methods to increase communal participation have become important techniques.

Wang's paper describes the evolution and characteristics of Chinese Ecological Agriculture. This developed to its current stage in the era of "family responsibility" that succeeded the Cultural Revolution in 1977. Prior to this date a great deal of ecological damage had been done as a result of the imposition of physical targets that took no account of ecological conditions. This damage continued after 1977 because of the production focus upon staple foods - mainly rice. A strategy in which development workers, "could not see the forest for the trees"

Ecological agriculture developed as a reaction to these tendencies. It is characterized by a focus on the following factors:
- intergration of sectors (eg.industry and agriculture), subsectors (eg.crops and animals) and functions (eg.input supply, production, processing)
- fitting cropping systems to the environment (niche planting) and using multiple enterprises (eg, tall and short plants, fish, ducks)
- integrated watershed management and planned sequences of planting for land reclamation.

In the early 1980s, Zhejiang Agricultural University, Hangzhou formed an interdisciplinary research group that now includes 30 teaching staff from 9 disciplines. This has led to the initiation of outreach programmes for integrated development, plus the establishment of a inter-disciplinary Agro-ecological Institute that is unique in China. Despite this progress, additional tools are needed to analyze areas such as farm-household production and consumption, common property resources, support services and community issues.

Mwangi's paper on the Kenya experience states that farming systems concepts were introduced to the country in 1975, but only became widespread after 1983. The Kenyan approach advocates the analysis of farmers' needs and priorities in the process of formulating regional and national development strategies. It also supports the partnership of research scientists, extension workers and farmers to improve the process of technology generation, so that appropriate technologies are developed.

The Kenyan concept of farming systems encompasses both technical and non-technical issues in helping small scale farmers. It also recognizes the importance of inter-institutional linkages, environmental considerations, communal participation and food systems.

INSTITUTIONALIZATION OF FSD AS PART OF RURAL DEVELOPMENT

by

B.N. de los Reyes[1]

1. INTRODUCTION

During the past three decades, there has been a continued evaluation of strategies and methodologies for rural development. One of these has been the Farming Systems Research (FSR). This developed because it was recognized that most of the results of researches in the past were not adopted by the farmers. They did not meet the needs and socio-economic conditions of the farmers.

FSR is a holistic and inter-disciplinary approach to solving the production problems of the small farmers. However, farming is complex in nature. Research institutions, on the other hand, are generally component and/or commodity oriented. After the initial stages of data gathering and diagnosis, problems are reduced to one or a few technological constraints, for which technological solutions are to be found. Little attention is given to the institutional and socio-economic constraints faced by the small farmers.

In 1983, FAO introduced Farming Systems Development (FSD) in its programme of work. This was developed to complement the activities of FSR. FSD stresses the sustainable development of resource-poor farmers; the analysis of the whole farm-household including aspects related to crop, recognition of the dynamics of agricultural development; the use of informal, inter-disciplinary and participatory methods. FSD includes the familiar stages of exploratory diagnosis, verification survey, analysis of constraints and opportunities, testing of interventions and dissemination.

FSD has been promoted in the Asia-Pacific Region through various meetings and consultations. Some countries have introduced the system in their programmes and projects with FAO assistance. Training has been conducted for people engaged in these projects. The replaceability and continuity of FSD in the region depends on how the activities are institutionalized.

---

[1] Regional Farm Management Economist, FAO Regional Office for Asia and the Pacific, Bangkok, Thailand.

2. FSR AND RURAL DEVELOPMENT

The Green Revolution of the 1960s brought about significant increases in food production to the point that most developing countries in the region have attained or are approaching food self-sufficiency levels. However, this barely benefitted the small farmers resulting in a widening gap between their income and level of living and the large farmers. Governments launched rural development projects to attain greater equity in the distribution of economic growth between urban and rural areas and also between economic/social classes in the rural areas, i.e. large and small farmers.

In the Asia-Pacific Region, most governments have formulated policies aimed at improving the conditions of the rural poor including the small farmers. In response to these policies, research institutions have focused attention on the needs of the small farmers especially in the less endowed areas of the country. The agricultural extension service have been trying to redirect their efforts towards the small farmers. Some countries have improved their training programmes for the small farmers while others have taken more drastic action of unifying extension and decentralizing services to make them responsive to regional and local needs. The support services system consisting of credit, marketing and input supply have also been improved to support the small farmers. These are the basic elements required for the successful implementation of FSD. The impact of FSD can therefore be maximized if the approach is made a part of the rural development programme.

On the other hand, FSD can make significant contributions in improving policies,k planning and implementing rural development programmes. For example the informal methods of collecting data, i.e. rapid rural appraisal can provide timely and up-to-date information which can be used as a basis for planning projects rather than waiting for results of large-scale surveys which are outdated before they become available. The methodologies used for delineating agro-ecological and farming systems zones will be useful in delineating areas for development and classifying farmers into relevant categories will be useful for planning and implementing target oriented projects while actively involving them in the process. FSD could also provide elements required in developing policies for rural development and improving support services.

3. METHODOLOGICAL ISSUES

Given the wide range of problems faced in FSD, further development and improvement of the methodologies used are essential. Dixon (1990) pointed out that few or no additional tools would be required for an expansion of farm system scopes to include livestock, horticulture, household production and consumption and off-farm work. However, the broadening of scope to the analysis of agricultural systems, which include farm-household production and consumption, common property resources and the local communities, support services and institutions is more challenging and would require additional tools, notably community systems analysis methods. Another likely development is a deeper understanding of household decision making and strategy formulation, and the development of parameters and tools for the analysis of dynamic problems including sustainability issues. The approach would also be applied to a wider range of development areas, including support services and policy analysis.

Substantial methodological work would be required for these purposes. These would include improvements in data gathering and analysis. FAO has developed the Farm Analysis Package (FARMAP) which has been used in some countries of the region. Its decided advantages over other packages is the use of standard codes for data collection and entry and the package can generate tables for analysis and presentation. Some users have found it difficult to implement and they have gone back to their own packages. One way to improve FARMAP is to make it easier to be integrated with other packages.

There are numerous methodological problems during on-farm and whole farm testing in FSD. Whereas FSR, for which procedures have been developed, has been mainly in testing of partial crop-technological solutions, FSD involves a combination of practices and enterprises. These practices and enterprises are interrelated, which means that a change in one will likely have affect on the others and the farm as a whole. For example, raising feed crops and livestock production are interrelated which will have effects on the use and distribution of farm and family resources such as land and capital. The decision of the farmer on the allocations of the different resources will have an effect on the income of the farm family. On-farm testing of new farming systems and whole farm testing including the effects of exogenous factors such as prices will require the development of long-term monitoring techniques.

4. INSTITUTIONALIZATION

The institutionalization of FSD is a complex process. It involves regularizing the different activities among many institutes and agencies which have specific mandates and long-established traditions which are difficult to change. An interdisciplinary team should be organized to help in developing the methodologies to be used and in planning appropriate projects around the specific mandates of the different research stations and institutions. A comprehensive application of FSD would require the adoption of a farming systems perspective in the different organizations.

4.1 Research

FSR has been a prominent feature of programmes conducted by different countries in the region with support from the international research institutes like

IRRI and from donor agencies like USAID, IDRC, World Bank, Rockefellow/Ford Foundation, etc. It initially started as cropping systems, dealing primarily with rice and gradually it expanded to other crops, livestock, fish and agro-forestry. Although the interrelationships between the different commodities may have been analyzed, this cannot be considered a whole farm approach.

Some countries in the region have established special institutes to be responsible for farming systems research. In Thailand, the Farming Systems Research Institute was established in the Department of Agriculture. The Farming Systems and Soils Resources Institute was established in the University of the Philippines at Los Banos. In Nepal, the Farming Systems Research and Extension Division was created under the Department of Agriculture. However, staff members are mostly from the physical and biological sciences working mainly on commodity (mostly crops) based projects. The socio-economic aspects of farming systems are neglected of if not very weak.

In order for these institutes to adopt a FSD perspective, additional staff members should be recruited from other sub-sectors of agriculture and socio-economic fields. They can interact with the agronomists in planning and implementing research projects so that the inter-relationships between commodities including costs and returns and their acceptability can be analyzed. In cases where it will not be possible to recruit additional staff, an inter-disciplinary team can be organized with members from other institutions to provide advice and guidance in planning and implementing projects.

4.2 Extension

The agricultural extension service is one of the biggest organizations in the Ministry of Agriculture concerned with transferring technology to the farmers. They can also play an important role in the development of technology for the small farmers. They live and work with the farmers and they are more knowledgeable about farmers' problems. The extension workers can assist in gathering data and information, arrange discussions, help articulate needs of farmers, and assist in conducting field trials and provide feedback to researchers.

The agricultural extension workers have been dealing so far with the introduction of a single practice/crop or a package of practices for a single crop. Under FSD, they have to use a holistic approach and assist farmers in making decisions on the selection and combination of enterprises to meet their objectives.

In order to institutionalize FSD in agricultural extension, policies should be formulated that will enable the extension workers to respond to the legitimate needs of the small farmers rather than implementing programmes with little or no relevance to their needs. The extension workers must understand the whole farm-household system. They should be trained in the farming systems development approach so they can effectively work with farmers and researchers. The extension workers must also be backstopped by a strong team of subject matter specialists with training in FSD.

The farming systems approach require the involvement and participation of the farmers. The extension workers need training in the behaviourial sciences so that they can motivate farmers to actively participate in meetings and discussions. They should also learn how to prepare extension messages and how to disseminate information through such techniques as farmer-train-farmer, use of mass media, etc.

4.3 Support Services

The adoption and impact of the new technology will only be minimal unless accompanied by changes in support services such as marketing, input supply and credit. The agencies involved in providing these services need to adopt a farming systems perspective so that they can provide the needs of the small farmers. For example, lending institutions have developed good credit programmes for monocrop farming systems. Under FSD, farmers will need credit for different farm enterprises and non-farm activities which are interrelated. This requires the development of an integrated financing scheme to provide all the credit needs of the farmers, both short term and long term. A knowledge of farming systems development will enable the credit officers to meet the various needs from gathering and analysis of data to preparing credit programmes and farm plans and budgets on which to base the credit needs and repayment schedule to the small farmers. In the Philippines, FAO assisted the Land Bank in training the staff in applying farming systems approach in developing credit programmes in the agrarian reform areas where farmers have become amortizing owners.

4.4 Agricultural Policy Analysis

The success of FSD depends on the commitment of the government and support provided by different institutions and organizations. Policies and programmes should be formulated to provide proper incentives for farmers to adopt improved technologies. A careful analysis of the farm-household and agricultural systems is important to understand the likely impact of agricultural policies on product prices and farmers' incomes. In order to do this, there should be a continuous flow of information from the field about farmers' resources, farmers' goals, production and consumption patterns, resource use, prices paid and received, etc. In Thailand, FAO assisted the Office of Agricultural Economics, Ministry of Agriculture and Cooperatives in collecting and analyzing household data for policy formulation. A similar project of the Agricultural Price Commission in India was supported by FAO to facilitate the analysis of data collected in the field for setting price support.

4.5 Education and Training

The availability of trained manpower is essential for the institutionalization of FSD. In addition to the in-service and short-term training by the institutes and organizations, there is a need to develop higher level training to provide people who can take the leadership in developing programmes, methodologies and implementing projects.

There are now a few colleges/universities offering farming systems as part of their curriculum. Chiang Mai University (CMU) and the Asian Institute of Technology, both located in Thailand, offer Master of Science degrees. The University of the Philippines at Los Banos (UPLB) and Khon Kan University (KKU) in Thailand are offering one-year post-graduate certificates and diplomas, respectively.

The core teaching programmes centre has an understanding of systems theory and analysis and their application in farming systems research. An important aspect of FSD is its broader application in the socio-economic analysis, policy formulation and farmers participation. The curriculum could be improved by strengthening the socio-economic aspects and the behaviourial sciences so that the graduates will have a better understanding and appreciation of the conditions of the small farmers, their aspirations and needs.

REFERENCES

Fresco, L.O. Issues in Farming Systems Research, Netherlands, Journal of Agricultural Sciences 32 (1984). 253-261.

Dixon, John. Ways Forward for the Farming System Approach in Asia and the Pacific, 1990 Asian Farming Systems Research and Extension Symposium, AIT, Thailand, (Mimeographed).

INSTITUTIONALIZING FARMING SYSTEMS RESEARCH AND EXTENSION IN THE PHILIPPINES

by

W. Dar[1]

1. INTRODUCTION

Farming systems as an approach to agriculture and rural development in the Philippines is widely accepted. Yet, there are two major factors preventing the full institutionalization of farming systems research and extension (FSRE) in the country. First, the major practitioners have not fully agreed on what should constitute the most appropriate framework for farming systems. Everyone agrees that the idea of systems is basic to the approach and yet, what elements should go into the system are not fully delineated. Everyone has also fully endorsed the use of an inter-disciplinary approach. However, records show that the team does not work as expected. For instance, most of the terminal reports are usually separated by discipline. Thus, there is a need to identify the parts or elements of FSRE, including their linkages and interrelationships.

Second, FSRE is heavily oriented towards the improvements that can be transferred to the farmers to increase farm productivity. Thus, research in agriculture focuses on the search for an improved technology that the farmers can adopt. In other words, there is a strong technology bias in the design of FSRE. This has led to a methodological approach oriented towards <u>diagnosing agricultural problems</u> with the goal of overcoming the constraints to production.

In view of this, the interaction between on-station and on-farm research is focused on the stages of developing technology, i.e., technology generation, adaptation, verification and dissemination. In spite of the systems claim of farming systems practitioners, the focus of research is <u>on the technology</u>, which is expected to completely replace the existing technology being used by the farmers, because the latter, apparently, does not increase production.

Such methodological orientation, however, has created more problems than solutions. It has effectively served as a blinder, so that we fail to see FSRE as an approach. A recent evaluation of FSRE in the Philippines has revealed that technology transfer in the country remains a mystery. This conclusion cannot be

---

[1] Director, Bureau of Agricultural Research, Department of Agriculture, Philippines

taken lightly. There must be reasons why this phenomenon prevails. The major one relates to the extensive emphasis of research on improved technology, in spite of the general recognition that there are other factors that would affect the transfer of technology, such as the socio-economic factors. However, often, the role of these factors is not woven into the technology transfer process.

We believe that unless we develop the skill in identifying the elements that would constitute the farm as a system and in building scenarios based on alternative interrelationships, institutionalizing FSRE will still face great difficulties.

At present, the linkage between research and extension (R-E) has not been fully operationalized, although there is a general agreement that the two must be linked. There is still the nagging problem of how to overcome the dictum of "where research ends, extension begins." Presently, the delivery of services is effected by disjoint efforts whereby the researchers conduct studies, the results of which are given to the development communicators who, in turn, package them for delivery and transfer to the farmers by the extension workers. But many research results cannot be transferred to the farmers, because they are not relevant to the latter's needs and problems.

Attempts to overcome this problem are being instituted. For instance, the farmers are encouraged to participate in the research process. Yet, there should be continuing consultation to understand farmers' needs to ensure their participation until the technology has been developed. Such understanding can be facilitated by the extensionists. But the farmers' and the extensionists' involvement in the research process still needs to be properly conceptualized.

The process of institutionalization is a mechanism by which those engaged in a given activity abide by the current rules and regulations explicitly stated or implicitly agreed upon by the members of social institutions. The agreement carries with it obligations and sanctions.

The question, then is this: can FSRE be institutionalized? In other words, can this approach develop a set of rigorously formulated procedures to which all practitioners will adhere? Our answer is definitely affirmative. However, this adherence to the rules and procedures will not happen overnight. Before this can occur, a process of cultivation by building up the needed shared experience among the practitioners is required. This body of experience will be used, in turn, to train the prospective recruits, who will then be called farming systems specialists.

2. SELECTED FSRE EXPERIENCES

A study of the experiences in FSRE in the Philippines can provide lessons which can be used to draw the requirements for the institutionalization of FSRE. Some of the experiences are given below.

2.1 Eastern Visayas Farming Systems Research Project (Cornick, et.al., 1985)

The Eastern Visayas Project had three collaborators: the then Ministry of Agriculture (MA), Visayas State College of Agriculture (VISCA) and Cornell University (CU). It began in 1982 with a high commitment to involving farmers in the development of small-scale upland farming technologies. The experiences in two villages are compared below.

In Barangbang, discussions had repeatedly identified soil fertility and erosion as problems. The researchers' suggestions for planting Leucaena on field contours as hedgerows and for fertility-enhancing cropping patterns were accepted in principle by the farmers, but they were reluctant to allocate scarce land to Leucaena until nine farmers had been taken on a visit to a neighbouring island (Cebu) to see how Leucaena was being used there.

The research agenda was finalized with the help of the village residents, and trials rapidly went ahead, limited only by seed availability. Enthusiastic cooperators planted not just the trials, but also on their other plots, including some of the control plots. This interfered with the proper conduct of the research trials.

The farmers sold the leaves to processors of animal feed, for a ready cash income, instead of incorporating them into the soil as the researchers had recommended. They also experimented with pruning practices, and with appropriate crops for shaded areas. They conducted "farmers teaching farmers" training courses and began to collaborate in the control of the leafhopper Psyllid, a source of damage on Leucaena.

The use of the farmers' association and incentives to maintain and reward farmer cooperation divided the community between those directly associated with the project and those without direct association. However, the participatory base remained and the continuing role played by the farmers was considered to be the most important feature.

On the other hand, in Jaro, research focused on the possibilities for improving the lengthy fallow period through the restoration of soil fertility, control of Imperata cylindrica, and reduction of costs for clearing and improving pastures.

Just like in Barangbang, meetings, surveys and discussions were held. The farmers chose the legume type and combination they wanted to test from different trial designs.

Problems with legume establishment and cooperation became evident after the first planting. The Pueria cuttings failed to establish and Centrosema had a low germination rate and slow growth. Additional trials were planted by the site staff. Many trials were damaged by grazing livestock. Six months later, they were ploughed under or abandoned.

In Barangbang, the technology was familiar, because the farmers had experience with local Leucaena varieties. It provided significant short-term cash benefits. In Jaro, the recommended techniques for legume establishment were untested and the farmers had no information about expected growth rates. Meetings for cooperators took place in the project office, biasing the discussion away from the possible dissent of non- cooperators. The researchers also took over the trials to safeguard the research program, rather than allowing time to do them at the farmers' pace. This undermined the participation of the farmers.

This shows that identifying the reasons for successful farmer participation is difficult, because the issues are confounded by the social structure and cultural norms. Cooperation may be part of appropriate farmer behaviour toward outsiders, superficially indicating agreement and involvement on the part of the farmers that perhaps do not exist. These problems are likely to be worse, if material incentives are offered.

2.2 Highland Agriculture Development Project (Fumucao, et al., 1987)

A Research and Extension Pilot Project was implemented from 1985-1987 as one of the activities of the HADP. The HADP aims to uplift the general well-being of the upland farmers and maintain the ecological balance in the uplands. The HADP covers Benguet and Mountain Province.

The pilot project was implemented in Barangay Camatagan, Sabangan, Mountain Province. It sought to introduce some of the recommended research and extension concepts of the HADP; identify the problems and issues in the introduction of these concepts; and recommend measures to resolve the identified problems in the implementation of the HADP research and extension agenda. A rapid rural appraisal (RRA) was done in the barangay. The RRA was followed by the drawing up of a research and extension plan. The activities focused on the improvement of the existing cropping pattern techniques in the area. This focus was identified based on the results of the baseline survey.

An interdisciplinary team composed of representatives from the different government institutions in the region was formed. The disciplines represented were animal science, agricultural research, home management, home management, agricultural extension, agronomy, agri-business, soil science and social science. This team was organized by the DA, through the Highland Agriculture Research Consortium (HARC), to manage the pilot project.

The team activities enabled each member to appreciate the disciplinal concern of the others. This understanding led to their holistic view of the village as embodied in the "Camatagan Farming System."

One of the highlights of the pilot project was the organization of the farmer-cooperators association. During the first year of implementation of the project, free inputs were given to the farmer-cooperators. However, during the second year, an agreement was made which stated that the project would provide the inputs during planting time. But after the harvest, the farmers would pay the inputs.

The association then gave inputs to the farmers who have not served as cooperators. This gave the farmers a feeling of responsibility which enabled them to help themselves and the other farmers even after the project had ended.

The evaluation of the project revealed the ecological and social trade-offs. For instance, the performance of the trials indicated the potentials of vegetable production and fish culture in improving the living standards of the farm household through increased food production and income. The income opportunities offered by vegetable production, as indicated by the net return on production, encouraged the farmers to plant vegetables. This, however, was not a signal to abandon the concern to improve the yield of the traditional rice varieties, which were needed to make rice wine. The high yielding varieties did not suit the "social" requirements.

Likewise, the intensification of vegetable production may lead to the appearance on new pests and diseases, or even an increase in the number of the existing ones.

2.3 "Kabusugan sa Kaunlaran" Project (Dar, 1990)

The "Kabusugan sa Kaunlaran" (Bounty in the Farm) project had KABSAKA as its acronym. It was an inter-agency effort in the province of Iloilo led by the then Ministry of Agriculture (MA), through the National Food and Agriculture Council (NFAC). The Provincial Government, International Rice Research Institute (IRRI) and the then Philippine Council for Agriculture and Resources Research (PCARR) cooperated with the MA.

KABSAKA aimed to raise farm incomes by increasing cropping intensities and production in the traditionally single-cropped rainfed areas of Iloilo. The main intervention was the introduction of a multiple cropping technology involving livestock and fish production as determined by the farmers' resources. The other components were the marketing and transport systems, seed supplies, human nutrition and health, institutional staff resources, construction of small water impounding systems, pilot village development and conduct of adaptive research on rainfed cropping.

The KABSAKA experience was an example of the impact of an acceptable technology coupled with community-oriented support services. The acceptability of the technology was ensured through vigorous on-farm trials involving the farmers.

Even without the expensive peripheral efforts, the project, by banking on the basic multiple cropping component, would have achieved much of its targets. The important thing was that the Iloilo farmers learned to appraise the merits of the technology on their own and used this according to their resources and capabilities. A strong indication that this, indeed, is the case was that two years after the end of the project, multiple cropping continued to spread beyond the original adopters.

2.4 Manjuyod Project (Granert, 1991)

The project is implemented by the Soil and Water Conservation Foundation (SWCF) using funds provided by the Negros Rehabilitation Development Foundation (NRDF) in Negros Oriental. It directly assists 150 farm families. The farmers are informally organized into 11 groups of 5 to 10 farmers each.

The farmers have built soil and water conservation structures on their farms. These include contour hedgerows, contour rockwalls with hedgerows and contour canals with hedgerows. Other structures include the bench terraces, contour plot beds, soil traps and check dams. The farmers also make compost. They use compost pits, compost baskets and trenches. BSWCF provides farm input assistance. These include seeds of crops and trees; cuttings of sweet potato, black pepper and <u>Gliricidia sepium</u>; fertilizers; and veterinary drugs.

The Foundation also distributed shovels, digging bars, pick mattock, garden hoes, water hoses, carabaos and fishfingerlings. Input assistance and distribution is done under a matching grant scheme. The farmers pay for the inputs and animals by producing the seedlings, which are then used for reforestation. Farming systems are promoted. Fish-azolla and multiple cropping systems are now practised.

Training are provided and cross-visits are sponsored. Work groups have been formed. There are 18 of these. Farm and home visits are continuing activities of the technicians. Staff development, and monitoring and evaluation are done continuously.

The project has also initiated the formation of Manjuyod 2001, multi-sectoral group of government and non-government organizations (NGOs) and interested individuals. The group is involved in the restoration and protection of the watershed. Some of the lessons learned under the project are:

1. The major criteria for site selection should be the presence of:
   a. good, peace and order situation; and
   b. diverse land capability and land use systems to create diversity needed for stability and to serve as models for various technologies.
2. There are several mechanisms to stimulate community participation. Among these are the use of indigenous community organizations to implement activities; introduction of income-generating projects; generation of revolving funds for capital; use of materials inputs as stimuli; and facilitation of work

group formation to implement activities.
3. There are several technologies which provide immediate returns and are environmentally sound. These groups of technologies are soil-water conservation and soil fertility enhancement; multi-storey cropping; livestock integration; communal reforestation; and upland aquaculture.
4. The project's mechanisms to promote the spread of the technologies enabled it to sustain the interest of the farmers. These mechanisms include the following: involvement of the local government from the beginning of the project; use of on-farm and farmer-run trials and demonstration; use of cross visits; provision of starter materials, such as seeds, tools, fertilizers and livestock; holding of training by other farmers and credible staff; production of appropriate extension materials; access to credit; and provision of work animals.

2.5 Community-Based Forest Research and Development (R & D) Project (Anon., 1991)

The project was started in 1984 by the then Central Visayas Forest Research Institute (FORI). It terminated in 1989. It aimed to identify forestry researches that are of economic importance to the rural communities; test and verify the effectiveness of the community capability building approach; improve the socio-economic condition of the participants; encourage self-reliance and sustainability of activities; and establish closer people- government partnership.

Its implementing strategies were community organizing, farm demonstrations, training, cross visitation, technical and financial assistance, and facilitation of the delivery of basic services by the other government agencies to the community.
It yielded the following lessons:

1. Government programs should not be "canned" and then delivered to the farmers. They should be sensitive and responsive to the people's needs.

2. A cohesive organization of farmers can be developed, if they are made aware of its advantages with the guidance of the extension worker.
3. A continuous social preparation should be done to sustain the people's mobilization, cooperation and participation.
4. Technology packages should be referred and disseminated to the clienteles.
5. The establishment of livelihood projects should be one of the implementing strategies in any upland development project.
6. Sustainability and self-reliance are encouraged through the formation of a strong community organization which can be gradually transformed into a cooperative.
7. An organization, with the help of the extensionists, can facilitate the access of the members to the services of the government and NGOs.

2.6 Research and Outreach Subproject

For the last four years, the Department of Agriculture (DA), through the Bureau of Agricultural Research (BAR), has been implementing the Research and Outreach Subproject (ROS) of the Accelerated Agricultural Production Project (AAPP). The ROS responds to the need to correct the many problems associated with the research-extension linkage. The BAR felt that the emphasis on research in the linkage has led to the passive role of extension, where the latter is viewed mainly as a receiver of technology to be transferred to the farmers.

The ROS uses a unified approach to FSRE. This approach evolved out of the lessons learned from the various experiences of the DA in implementing programs and projects. The BAR, as the Project Implementing Unit (PIU) of the ROS, analyzed the frameworks and experiences of these programs and projects with the regional offices. Based on these documented lessons, the BAR reversed the order of the research and extension linkage. This led to the development of an extension-led research framework which recognizes that the farmers are the ultimate users of the results of agricultural research and development (R and D) programs.

This framework views non-adoption as caused by factors beyond what is commonly alleged as the farmers' traditionalism or ignorance. The ROS framework looks into the deficiencies of the technology as well as the process which generated it. The reversal of the learning process provides an opportunity for the researchers and the extensionists to learn from the farmers. Likewise, the location where research and extension takes place and the roles of the researchers and the extensionists are reversed or modified. Based on these features, the ROS framework adopted eight working principles, namely:

1. Market orientation - ensures adequate demand for farm products at a competitive price which is profitable to the farmers;
2. Farming systems approach - achieves farm family development through the complementation of on-farm, off-farm and non-farm activities. This is a function of the farm family's allocation of resources to three major production activities: crop production, livestock production and off-farm/non-farm activities;
3. Resource management orientation - considers the wide variety of resources found in the community, which should be mobilized to increase the options of the farmers and enhance their efficient use of these resources;
4. Community-based perspective - recognizes the existing social structure of production within the community and its influence on the farm management capabilities of each farm household;
5. Participatory approach - emphasizes real participation of all sectors concerned, especially the farmers, in all the phases of technology development, namely: technology identification, design or experimentation, dissemination, commercialization and evaluation;

6. Interdisciplinary orientation - recognizes the importance of the interactive roles of the various disciplines. This is needed due to the complexity and diversity of resource-limited farms;
7. Partnership -recognizes the importance of complementation and resource-sharing activities among the government organizations (GOs), both internally and externally; and between the GOs and the NGOs; and
8. Decentralized approach - allows the regional and provincial offices to exercise greater autonomy in the management of agricultural and rural development projects.

While ROS recognizes the constraints to production, it starts from the current opportunities and potentials of farm families. The aim is to improve their life by using the existing resources of the community. This means that research (R) and extension (E) begins from "where the farmers are and with what they have." Thus, the point of entry is always focused on improvement.

Unlike the conventional R and E process, the ROS extension- led research (E/R) focuses on improving current farming systems. It is not only the extensionist but also the researcher who links with the farm family. A unified approach is implemented and the interaction among the farm family, researchers and extensionists is maximized. The centrality of the farm family is a prime consideration, such that its involvement is evident in all the stages of FSRE. Research and extension begin and end with the farmer. Because of this, rapid rural appraisal (RRA) becomes the critical first step in FSRE.

Through the RRA, the current constraints to production, and opportunities and potentials for agricultural development in a given farm community, are identified. With the use of the ROS framework, the community-based plan is jointly evolved by the community, extension workers and researchers. This plan identifies the existing resources and other potentials for development in the community. The actors also plan the various strategies needed to mobilize these resources.

For instance, the initial activity can be the improvement of an existing production system, specifically, the cultural management practices. The elements include, among others, improvement of erosion control, pest control practices and postharvest technology management. In summary, the ROS process can be characterized as:

1. oriented towards research results and understanding the weaknesses of any given technology;
2. geared towards the enabling the researchers, extensionists and farm family to increase the relevance of E and R activities to farm management; and
3. addressed towards the improvements and innovations that bring about the desired changes in the community.

3. REQUIREMENTS FOR THE INSTITUTIONALIZATION OF FSRE

The <u>first requirement</u> for the institutionalization of FSRE is the <u>generation and adoption of a common perspective formulated</u> by an <u>Inter-disciplinary Team</u>. The perspective must be data-based. The varieties of "lessons learned" from the field must serve as the foundation of the framework. Thus, the various international experiences in implementing FSRE should be reviewed with the primary aim of synthesizing the "lessons learned," so that a general framework for farming systems development can be formulated. Therefore, there must be a <u>common conceptual framework for undertaking FSRE</u>.

To develop a <u>unified approach</u> to FSRE, the <u>interaction and complementation among the farm, the farmer/farm family and the farm environment should be taken as the organizing principle</u>. The farm is where production activities take place, and the family undertakes these activities. The farm environment consists of the forces that influence the family and its farm activities. It must be recognized that while the three elements are interactive and complementary, the focal point of orientation is the farmer/farm family. This must always be underscored.

In our current technology transfer activities, our focal point is the technology. Once an "improved" technology has been generated by research, it is assumed that all that needs to be done is to train and influence the farmers into accepting the technology. Our own field technicians always tell us that all we need to do is to <u>change the attitudes of the farmers</u>. How easy and how convenient for us. For in this situation, the farmers are blamed for the failure of technology transfer. The technicians have already done their best to teach the farmers and yet the latter resists the technology, because of their attitudes. This perception arises out of the wrong emphasis on technology transfer.

On the other hand, in the unified approach, since the focal point in the interaction and complementation is the farmer/farm family, the individuals must be taken as social psychological beings with values, attitudes, goals, needs and problems. Thus, what must be fully understood is the person's <u>social nature</u>. His nature is organized and shaped by his general social experience. His values and attitudes are influenced and shaped by his dominant relationship in the community. This fact is neglected in the design of FSRE. The use of cooperators and, at times, the use of a random sample of cooperators, violates the fundamental fact that man is, by nature, "social." Thus, the design of FSRE has to begin from the basic idea that there is a <u>social organization of production</u>.

The impact of technology transfer must always be seen from the perspective of the social organization of production and <u>not from the point of view of the individual</u>. Thus, there are five major dimensions that must be considered in developing a unified approach. These are: technical, social, economic, political and environmental. These dimensions are interactive and complementary.

Two examples can make this point clearer. First, it is well-known that the "better" farmers are more innovative because of their education and media exposure, among others. To hasten technology transfer, the "non-adaptors" must be trained and provided with more access to information. While these findings are conclusive, what is often missed in the analysis is the role of political factors. The better farmers' higher education and more exposure to the media is largely influenced by their political position in and outside of the community. Their strategic location in the community gives them more access to agricultural services and resources. Hence, they become better farmers.

Second, environmental protection is often related to and influenced by adoption. We are now at the stage where a new issue confronts agricultural research and extension. This issue is sustainable agriculture. Our fundamental goal is to make the environment productive and, at the same time, protect the environment. We are all aware that, to a large extent, environmental degradation is largely influenced by socio-economic and political factors. To some extent, some of those who are expected to protect our environment are directly involved in forest denudation. This is true most especially in the case of the logging industry.

We are also underlining the role of the Inter-disciplinary Team. While on-station research can readily establish the scientific validity of a given technology, its uses must be examined at the social level. Many studies have been conducted which revealed that, often the farmers would only choose from a package of technologies those elements that would fit their system. This finding requires a very careful analysis by an Inter-disciplinary Team.

For instance, there is a strong bias against farmers' tradition in FSRE. We believe that any planned intervention to be instituted in agriculture <u>must always begin from this tradition</u>. Research in agriculture must, therefore, begin by carefully <u>documenting and studying the current patterns</u> of agricultural production used in the community. We have to know those who have access to these technologies and why others have no access to them.

This issue brings us directly to the problem of the social stratification of farmers in the community. Clearly, equity and justice cannot be attained merely on the basis of providing those with less with more services. We have to reorganize the structure of social stratification in the community, so that those with less will have more.

Access to resources and services is largely affected by those who are in control of these resources. How this control is being exercised in the community requires careful documentation. Again, research data reveal that, on the whole, farmers are <u>rational</u> in their management of the farm. What really prevents them from using this rationality to the fullest is their lack of access to resources and services. This rationality is developed and supported by a body of traditions derived mainly from the entire range of farm management skills used by the farmers in their production activities.

In this light, the role of social scientists in FSRE becomes completely indispensable. The social scientist must no longer be considered as a mere appendage.

The <u>second requirement</u> is that <u>all researches</u> in <u>agriculture and rural development should adhere to the framework and adopt it in the development of research programs and projects.</u> Research should generate data that will continuously refine the conceptual framework. In other words, the unique experiences of each participating country must be used to adapt the framework to the requirements of the local conditions. Nevertheless, the major orientation of the framework must be retained.

It is also indispensable that the design of the extension system must adhere to the framework. In fact, this is what we are presently doing in our work. We are now in the process of institutionalizing our research and outreach framework in the regional and provincial operations of the DA.

This is very evident in our design of the National Agricultural Research and Extension Agenda (NAREA). The NAREA embodies the priorities in agricultural research and extension, which were identified through multi-sectoral and multi-disciplinary regional consultations. It aims to concentrate research and extension programs where they are most needed to maximize the use of scarce national resources.

Under the Agenda, agricultural research and extension activities are organized into four categories, namely: 1) priority development zones; 2) priority sectors under each development zones; 3) priority commodities by sector under each development zone; and 4) priority research areas for each commodity by sector under each development zone. These priorities are further stratified into three levels of activity, viz, technology generation (TG), technology adaptation or technology verification (TA/TV), and technology dissemination (TD).

We are reviewing the "lessons learned" in implementing our NAREA program. These "lessons" will be used to further increase the utility of the research results to agricultural programs and projects.

The <u>third requirement</u> is the <u>enhancement of the interaction and complementation of on-station and on-farm researches</u>. Our own efforts in this area are essentially focused on the attainment of consistency of focus, thrusts, and operational strategies to enhance the overall effectiveness of research station functions. In our program of rationalization, we are placing increased emphasis on the interaction of the researchers, farm family and extension personnel. In the development of our station activities we strongly consider the feedback from the farm family. We are also highlighting active multi-sectoral participation in our research and extension activities.

The fourth requirement focuses on the development of a strong FSRE network. We have already formalized our networking with state colleges and universities (SCUs). The network aims to: 1) serve as a mechanism for the exchange of experiences in institutionalizing the farming systems approach in the various ecosystems; 2) facilitate the exchange of expertise, methodologies and results of on-farm experimentation; and 3) foster inter-agency coordination and collaborative undertaking in serving the needs of the countryside. In our networking program, the SCUs serve as critics of our farming systems activities and provide us with new knowledge, especially those they would consider as "breakthroughs."

The fifth requirement is the periodic review of researches vital to farming systems development. In this connection, we are presently undertaking the Annual In-House Research Review. We also actively participate in the Regional Research Symposia of the regional consortia. The two have strong multi-sectoral participation. The basic aim of the reviews and symposia is to ensure that the various researches adhere to the canon of scientific procedure. The use of the results is ensured, because many of the possible users of the results actively participate in the activities.

The best results identified in the review are presented in the regional symposia to initiate the process of packaging these results into usable technologies for widespread and commercialization.

The sixth requirement is the installation of an integrated data management system in agriculture and rural development. The current state of data fragmentation must be overcome. Research and extension must work toward the generation of a unified body of research data. In view of this, there is a need to be always conscious of integrating two general types of data in agriculture and rural development. These are: descriptive and explanatory data. The former must concentrate on the systematic description of the farm as a system and the latter must focus on explaining the dynamics of the farm as a system.

The active role of the Inter-disciplinary Team must be central to the development of the data management system. The members of the Team must delineate the adequacy or inadequacy of existing data and reveal the current data gaps. An integrated approach to data management will be vital to multi-level decision-making in farming systems development.

The seventh requirement is the conduct of periodic workshops and symposia on the current issues and trends on FSRE. The utility of our framework should be subjected to periodic review based on new insights gained from new research data and on the current field experiences of our extension personnel, including those coming from the private sector.

The eighth requirement is training, including the design of fellowship and scholarship programs that will enhance varieties of expertise on FSRE. There are two levels of training: non-formal and formal. It is important that all training designs in agriculture and rural development be developed using a farming systems development framework. In other words, however limited and specific the focus of a training activity may be, it must, in all cases, be designed with an appropriate farming systems framework.

At present, many colleges and universities are offering programs in agriculture which are very much commodity-oriented. These programs must now be properly situated within the context of systems and relationships. In other words, in all the major fields of agriculture, all activities must be oriented toward the farm as a system. There is a need to develop a set of core courses which are focused on understanding and analyzing the farm as a system. While the major program departments should continue their own training, research and extension activities, the SCUs must create an appropriate inter-disciplinary structure that will ensure the coordination and complementation of curricular offerings under the umbrella of farming systems. Countries and institutions with extensive experience in FSRE must be utilized as training ground for those with inadequate experience.

The ninth requirement is the encouragement and support for the packaging and production of development communication materials committed to the development of a unified approach to agriculture and rural development. The major focus of these materials must be on the "lessons learned" in designing and implementing FSRE. They should view FSRE as a system.

The communication materials to be developed must be based on the format preferences and information needs of the end-users. Likewise, they should not be limited to the traditional tri-media materials. The potentials of indigenous and non-conventional media are waiting to be tapped fully.

CONCLUSION

The foregoing discussions have delineated the major requirements for institutionalizing FSRE. The need to develop an
appropriate framework was underscored. No amount of effort will succeed in institutionalizing the approach if a common perspective is not shared by all farming systems practitioners. In all instances of institutionalization, the sharing of a perspective is always an imperative. Institutionalization will not take root without a shared perspective. The major problems associated with inadequate data analysis and interpretation in agricultural research will be a thing of the past only when an appropriate perspective is shared.

The mechanisms for institutionalizing FSRE are those which are commonly used in development work. Most of the existing structures can be used, provided that a _systems perspective_ is used in every case.

Although our discussion of the role of the SUCs in institutionalizing FSRE is limited, it is now important that they embrace the _systems perspective_ in training the new cadre of experts in agriculture and rural development. In the final analysis, the educational institutions are mandated to socialize the new cadre of experts into the requirements of FSRE institutionalization.

REFERENCES

ANON. 1991. Highlights of Upland R and D Program Activities of DENR Region VII. Paper presented at the Workshop on R and D Focused and Upland Areas Requiring Conservation, DA-Region VII, Cebu City, January 16-17.

BAYACA, R. 1990. "Promoting Farming Systems Development Through Increased Farmers Participation: The AAPP-ROS Experience. "In Dar W. D. (ed.) Basic Elements of the Farming Systems Approach to Research and Extension. Diliman, Quezon City: Department of Agriculture, Bureau of Agricultural Research.

BEDEIAN, A. G. 1980. Organizations: Theory and Analysis. (2nd Edition). Hindsdale, Illinois: The Dryden Press.

BONIFACIO, M. F. 1988. Working Papers on Community-Based Agriculture. Quezon City: Agricultural Training Institute.

_____. 1991. A Critical Assessment of Research-Extension Linkage: Toward a Framework for Technology Transfer (A Partial Report submitted to the Socio-Economics Research Division, PCARRD).

BUREAU OF AGRICULTURAL RESEARCH. 1989. The National Agricultural Research and Extension Agenda 1988-1992. Diliman, Quezon City:Department of Agriculture, Bureau of Agricultural Research.

BYERLEE, D., P. HEISEY and P. HOBBS. 1989. "Diagnosing Research Priorities for Small Farmers: Experience from On-Farm Research in Pakistan. "Quarterly Journal of International Agriculture. Vol. 28, N. 3/4.

CHARRY, A. and J. L. DILLON. 1989. "Structuring National Research with a Farming Systems Perspective of the Tropical Savannas of Columbia. "Quarterly Journal of International Agriculture. Vol. 23, N. 3/4.

CORNICK, T., D. ALCOBER., R. REPULDA and R. BALINA. 1985. "Farmer Participation in OFR and E: Some Farmers Still Say "No:" Lessons from the FSDP Eastern Visayas." Paper presented at the Farming Systems Research Symposium, Kansas.

DAR, W. 1990. Experiences and Directions of Farming Systems Research and Development in the Philippines. Paper presented at the 12th Annual Scientific Meeting of the National Academy of Science and Technology, PICC, Manila, July 11.

DOPPLER, W. 1989. "Current Approaches and Future Potential of Farming Systems Research." Quarterly Journal of International Agriculture. Vol. 28, N. 3/4.

DY, M. E.Y. 1987. Communication Models, Theories and Strategies. Batac, Ilocos Norte: Mariano Marcos State University.

FLORA, C. BUTLER and M.TON CEK. (eds.). 1985. Farming Systems Research and Extension: Management and Methodology. Paper Abstracts N.10, October, 1985. Manhattan, Kansas 66506 USA: Kansas State University.

FUMUCAO, M., C. WANGDALI, ET AL. Final Report of the Research and Extension Pilot Project Component of the Highland Agriculture Development Project.

GRANERT, A. 1991. Highlights and Accomplishments of the Soil and Water Conservation Foundation Projects. Paper presented at the Workshop on R and D Focused on Upland Areas Requiring Conservation, DA-Region VII, Cebu City, January 16-17.

HAEN, H., DE and A. RUNGE-METZGER. 1989. "Improvements in Efficiency and Sustainability of Traditional Land Use Systems through Learning from Farmers' Practice." Quarterly Journal of International Agriculture. Vol. 28, N. 3/4.

HAYAMI, Y. 1981. "Agricultural Development: The Japanese Experience." In: Drilon, J.D., Jr. and G.F. Saguiguit. (eds.) Accelerating Agricultural Development. College, Laguna: SEARCA.

HEIDHUES, F. 1989. "Introduction: The Farming Systems Concept in Agricultural Research." Quarterly Journal of International Agriculture. Vol. 28, N. 3/4.

HILDEBRAND, P. E. and F. POEY. 1985. Evaluating On-Farm Research." Farming Systems Support Project Newsletter. Vol. 3,N. 3. Third Quarter.

HORTON, D. and G. PRAIN. 1989. "Beyond FSR: New Challenges for Social Scientists in Agricultural R and D. "Quarterly Journal of International Agriculture. Vol. 28, N. 3/4.

HUNTER, N. D. and T. FARNINGTON. 1985. "Farming Systems Research in Botswana." Farming Systems Support Project Newsletter. Vol. 3,N. 3rd Quarter.

LABIOS, R. V. 1990. "The Farming Systems Research and Extension Process." In Dar, W. D. (ed.). Basic Elements of the Farming Systems Approach to Research and Extension. Diliman, Quezon City: Department of Agriculture, Bureau of Agricultural Research.

_____. 1990. "Farming Systems Approach To Research and Extension: An Experience of PHARLAP (Philippine- Australian Rainfed Lowland Antique Project)." In Dar, W. D. (ed.). Basic Elements of the Farming Systems Approach to Research and Extension. Diliman, Quezon City: Department of Agriculture, Bureau of Agricultural Research.

_____, and H. C. HEREZA. 1990. "Farming Systems Approach to Research and Extension: An Experience of the Capiz Settlement Project." In Dar, William D. (ed.) Basic Elements of the Farming Systems Approach to Research and Extension, Diliman, Quezon City: Department of Agriculture, Bureau of Agricultural Research.

MATHEMA, S.B.,D.L. GALT, ET AL. 1986. Report on the Process of the Group Survey and On-Farm Trial Design Activity, Naldung Village Panchayat, Kavre District, Nepal, Department of Agriculture Socioeconomic, Research and Extension Division.

MERRIL-SANDS, D. 1985. "A Review of Farming Systems Research." Paper Prepared for Technical Advisory Committee/CGIAR.

EWELL, P., S. BIGGS, AND J. McALLISTER. 1989. "Issues in Institutionalizing On-Farm Client-Oriented Research: A Review of Experiences from Nine National Agricultural Research Systems." Quarterly Journal of International Agriculture. Vol. 28, N. 3/4.

OSBURN, D. D. and K.C. SCHNEEBERGER. 1978. Modern Agricultural Management. Reston, Virginia: Reston Publishing Inc., A Prentice-Hall Company.

PICKERING, D. C. 1985. "Agricultural Research and Extension in Francophone West Africa." Farming Systems Support Project Newsletter. Vol. 3, N.4, Fourth Quarter.

POATS S., D. GALT, ET AL. 1987. "Farming Systems Research and Extension: Status and Potential in Low-Resource Agriculture ."Farming Systems Support Project Newsletter. Vol. 5,N. 3, Third Quarter.

_____. 1987. "Future Directions for FSR/E." Farming Systems Support Project Newsletter. Vol. 5, N. 3, Third Quarter.

_____. 1987. "Relationship Between FSR/E and Single Commodity Research Programs." Farming Systems Support Project Newsletter. Vol. 5, N. 3rd Quarter.

RIGGS, F., A. GOMEZ, ET AL. 1989. Rainfed Resources Development Project: An Evaluation Report.

ROSARIO, E.L. 1990. "Farming Systems As An Approach to Agriculture and Rural Development." In Dar, W. D. (ed.).Basic Elements of the Farming Systems Approach to Research and Extension. Diliman, Quezon City: Department of Agriculture, Bureau of Agricultural Research.

WEINSCHENCK, G. 1989. "From Subsistence Households to Sustainable Farming Environment Systems." Quarterly Journal of International Agriculture. Vol. 28, N. 3/4.

AGRO-ECOSYSTEM AND FARMING SYSTEM DEVELOPMENT, THEIR SIGNIFICANCE TO AGRICULTURAL DEVELOPMENT IN CHINA

by
Z. Wang[1]

1. FROM CROPPING SYSTEMS TO ECOLOGICAL AGRICULTURE

Since the 1950s the development of agriculture in China has passed through several phases. Firstly, priority was given to component technologies in the context of cropping systems development, such as the improvement of variety, fertilization and plant protection, so that production could be restored. At the same time, cropping intensity was increased: in northern China, from 100 to 150 percent; while in southern China, the single rice crop was replaced by triple-cropping which required the use of high-yielding varieties, diversified rice nursery methods and improved fertilization, control of pests and diseases, and construction of paddy fields.

Subsequently, Zhejiang Province, which had the highest grain yield, formulated an approach that required the integrated management of agriculture, forestry, animal production, side-lines and fisheries and the rational planning of mountains, rivers, farmland, forests and roads. Agricultural production greatly increased until the Cultural Revolution, which promoted "reclamation up to the peak of mountains and planting of rice to the centre of lakes". As a result, natural resources including forests were destroyed; and animal production, fishery and side-lines and consequently farmers' incomes, declined and "poor villages of high yield" became common.

After the Cultural Revolution, the individual household responsibility system was introduced. Farmers' incentives increased dramatically. However,

---

[1] Wang Zhaoqian, Professor, Director of Agroecology Institute, Zhejiang Agricultural University, Hangzhou, Zhejiang, China.

the exploitation of forests and grasslands by farmers continued, which was aggravated by population increases and decreases of arable land. Since 1984 aggregate agricultural production has not expanded and the destruction of ecological equilibrium became apparent. On the other hand, many farm households successfully integrated the production and processing of crops, forests, animals and fish, by managing their farming systems according to ecological and economic principles. Many noted the resemblance between such Chinese ecological agriculture and integrated farming systems, since the goals of both are the full use of resources and the conservation of the environment, consistent with the approaches to sustainable agriculture in foreign countries.

Historically, early research on farming systems in China was influenced by the ideas on grassland and farmland rotation of the Russian soils scientist and agronomist Williams, who placed emphasis on the improvement of soils as the basis of integrated agricultural management. The role of animal production and forest windbreaks for insuring the overall prosperity and sustainable development of agriculture were also emphasized. This approach, coupled with Chinese traditional experience, led to a first peak in the perception and understanding of farming systems in the Chinese academic world. After the Cultural Revolution, there was more Western influence on academic thought, notably the brothers Odum (concepts and methods of ecology) and Spedding, Conway and DeWit (systems views and research methodologies). The integration of Western systems views, research methods and economics, together with the lessons of failure in China, led to the theoretical foundations for the holistic management of ecological agriculture. Now, "Agricultural Ecology" is offered as a course in several Chinese agricultural universities.

2. THE SYSTEMS CONCEPT: THE INEVITABILITY OF HISTORICAL DEVELOPMENT

Since the 1970s, natural scientists (A. Toffler, I. Prigogine and H. Haken et al) noted the emphasis in western civilization on reductionism, i.e. the separation of a problem into parts as small as possible. However, it was often forgotten to combine the knowledge gained of parts into a holistic understanding. Traditional Western science emphasized stability, order, equilibrium and linear relationships. Failure is inevitable if an open system, that exchanges materials, energy and information with its environment, is analyzed mechanistically. The transition from industrial society based on large intensive use of energy, capital and labour to a highly developed society featuring information, invention and self-organization of systems requires a new scientific model. The emergence of new concepts and scientific approaches under such circumstances is historically inevitable.

Nationwide resource degradation, such as deforestation, deteriorating grassland and the reclamation of lakes for grain production, resulted in reduced grain, animal and fish production, soil and water losses and droughts and flooding. For example, approximately 1.6 billion ton of soil, which contains about 40 million ton of N, P and K, is eroded annually in the northwest plateau of China reaching 150t/ha in the most seriously eroded areas. Approximately one quarter of this material accumulates in the bed of the lower reaches of the Yellow River, resulting in an average annual increase in elevation of the bed of about 1 cm. Serious flooding is inevitable, as witnessed during the past two years.

Intensification requires careful assessment of all aspects of the system. Research into the promotion of high yields (15t/ha/yr) of grain in Shaoxing County, Zhejiang Province showed that under high levels of fertilization (750 kg/ha of N P, and K) the marginal productivity of K was relatively high (6.4 kg grain 1 kg of K) but the leaching of nitric acid after application of fertilizer increased rapidly, resulting in serious pollution. With regard to other aspects of soil fertility, the organic matter content of the rich black soil in the plains of the Heiliongjiang Province decreased from 6 - 9% before the 1940s to 3 - 5% now. Soil organic matter in the plains of northern China is generally below 1%. In the high-yield areas along the middle and lower part of the Yangtze River the soil organic matter has decreased to below 2%.

Imbalances between agriculture and industry are illustrated by two examples. Brickfields operated by farmers resulted in fluoride and dusts air pollution and thus the loss of cocoon production in Zhejiang Province reached 3850 tons in 1984. The cultivation of yellow peach trees was encouraged on hill land in one county. However, this variety of peaches was not suitable for fresh consumption, and the limited processing capacity could not absorb all of the harvest.

The above-mentioned examples demonstrate that agriculture, as a highly complex system, cannot be sustainably managed by reductionist methods and that the system also interacts in vital ways with its physical and socio-economic surroundings.

3. PRACTICE OF CHINESE ECOLOGICAL AGRICULTURE

3.1 Distinction between Chinese Ecological Agriculture and Western Ecological Agriculture

Chinese ecological agriculture has developed according to local conditions, *viz*, a large population, little arable land and low levels of other resources per capita. Arable land area per capita in China is only one quarter of the world's average (0.37 ha), and about one third of the arable land has production problems. However, China has abundant labour resources and a tradition of intensive cultivation.

Therefore, Chinese ecological agriculture emphasizes an high input use, intensive cultivation, together with the conservation of resources and environment. It also emphasizes recycling energy and materials and their transformation in a chain, and considers ecological as well as economic and social efficiencies.

3.2 Applications of Ecological and Economic Principles

Many ecological principles are applied in the sound management of agroecosystems, including: energy transformation and agricultural potential based on the primary productivity of the biosphere; harmony between organisms and their environment; carrying capacity of resources, the laws of limiting factors and tolerance; food chains and food webs; niche, time and spatial structure; material cycling; and stable increases of productivity; systems ecology based on systems and synergetic theory; co-evolution and ecological succession in agroecosystems; and diversity of species.

Therefore, the effective management of agroecosystems and the development of holistic, balanced integrated agriculture requires research on energy transformation, material transfer, efficiency and flow of information accompanying flows of energy and materials so that the systems productivity, stability, sustainability and resilience will be enhanced. Furthermore, methods of systems engineering and mathematical modelling facilitate quantitative research on agroecosystems. From the economics perspective, principles regarding the rational combination of resources and land utilization, farmer decision-making and market behaviour are also applied in ecological agriculture. The intended results are practical integrated complexes of ecosystem, socio-economic and technical systems, which effectively serve farmers and national needs.

Traditional Chinese agriculture contains abundant simple ecological and often successful experiences which rely on intensive cultivation and traditional organic farming to conserve soil productivity. Chinese ecological agriculture is a result of integration between traditional experiences and modern scientific knowledge, and can be defined as an agroecosystem with high ecological, economic and social efficiencies attained under the guidance of ecological and economic principles through the adoption of methods of systems engineering, holistic and appropriate agricultural production structures and integrated modern agricultural techniques.

3.3 Diversified Patterns Adapted to Various Areas

Ecological agriculture has vitality because its theory has universal significance for diverse, flexible and adaptive farming systems. Presently, the major production patterns practised in China have the following significant aspects:

a) <u>Integrated Enterprises</u>: Integration of crop and animal production and processing industry. Integration refers to both economic and ecological relationships.

b) <u>Diversified Cultivation</u>: Integration between environment and organisms and the practical application of niche theory to properly arrange for the primary production by a rational arrangement of crops planted on suitable lands. For example, the ratio of grain crops to fruit and vegetable crops in Dong-yuan County, Guangdong Province improved from 72:28 in 1978, to 53:47 by 1990.

c) <u>Sequences</u>: A special shrub was used as a pioneer plant on badly eroded steep land in Wushan County, Sichuan Province, followed by the establishment of Eulaliopsis binata, which contains soil erosion and is also profitable. Sometimes, small adjustments in planting methods can improve markedly the ecological productivity. For example, Heiliongjiang Province used to cultivate seedings to enable earlier planting.

c) <u>Vertical Utilization</u>: Vertical utilization involves integrated management of the upper slopes and the associated lower parts of the ecosystems. Generally, mountainous land can be divided into three vertical zones: the upper (protection) zone for trees or grasses; the middle zone of hilly area for integrated agriculture, forestry and animal production; the lower zone of flat land for high-yielding cereals combined with other crops and animal production. The three zones are interdependent. In order to conserve high mountains and steep slopes, there must be enough food produced on the low plains, and the middle zone should contribute to the farmer's income and afford protection to the upper zone. Similarly, individual fields can be vertically utilized based on traditional diverse intercropping and relay cropping. For example, shade-tolerant crops or fodder can be planted under grapes. Sometimes, deep ditches between grape rows provide drainage and also the production of fish, shrimps, and crabs.

d) <u>Complementarities</u>: Mutually beneficial and complementary systems could be established either by mimicking natural ecosystems formed through ages of evolution and succession, or by organizing various intercropping and/or multiple cropping patterns through a selection of mutually beneficial crops. The artificial communities of "rubber plus tea" and "rubber plus tea and medicinal plants" in Xishuanbanna, Yunnan Province are examples of such patterns. In Baoji City, Shangxi Province, the integrated utilization of above and below ground field and homeyards was proposed and patterns of "fruit plus grain plus fertilizer", fruit/mulberry plus poultry and fish", "forest plus medicine and mushrooms", "forest plus grain and grains", "fish plus pig and mulberry", and "grain plus vegetables" etc. were developed.

e) <u>Symbiosis:</u> About half of paddy fields in Mei County, Sichuan Province contain fish, generating a combined net return of double that from rice alone. Rice has also been cultivated with snails, crabs, taro and azolla. The azolla and taro leaves served as feed for snails, crabs and fish, added organic matter in the fields and provided shading for fish in summer. Annual income increased by 150% compared to sole rice.

f) <u>Agroforestry:</u> In northern China, agroforestry systems mainly take the forms of crops planted between rows or within a network of tree windbreaks. In southern China, a wide variety of woody plants such as arbores, fruit trees, mulberry trees, tea, bamboo, are integrated with crops to produce diverse agroforestry systems with ecological and economical linkages. Trees, such as chestnut and apple are intercropped with shorter-lived peach trees, and inter-planted with wheat or legumes (e.g. in Qianan County, Hebei Province).

g) <u>Ecotones:</u> The transition areas between land and water body between mountainous land level ground or plain, and even between desert and farmland are ecotones. These are fragile and unstable areas, but in the northern plain of Zhejiang Province, patches of upland, lowland and ponds, production patterns composed of grain crops, animals, fish and mulberry have existed for more than five hundred years. In recent years, duck and fish production along the sides of ponds and streams has succeeded and greatly expanded in area.

h) <u>Food Chains:</u> The pattern of "chicken-pig, biogas-fish" (sometimes with mushrooms) was widespread in southern China. This pattern was combined also with crop production, especially grains. In northern China, seven food chain production patterns, notably: "pig-biogas, grain-fruit" "pig-biogas-grain-vegetable-melon-mushroom", "beef cattle-biogas-pig-grain-grass" etc. were extended in Ankang Region, Shanxi Province. The income per capita reached 1080 yuan in 1990, respectively.

The above patterns should in all cases be adapted to local conditions, and to evolving economic conditions. Ecological agriculture should not only pay attention to cultivated land but also to the natural and social resources of the whole local system. It is necessary to consider agriculture, forestry, animal production, side-line and fisheries arranging the production structure to achieve an agroecosystem and local social system equilibrium.

3.4 Steps and Methods of Practice of Chinese Ecological Agriculture

Zhejiang Agricultural University (ZAU) has outreach sites in hilly areas of the Nanfeng Township of Jiande County, Zhuma Township of Jinghua City, and Shanqiao Township of Deqing County. The University is now testing a standard methodology, beginning with survey, zoning, diagnosis, planning, designing, evaluation, monitoring, adaptation, etc. In planning for the Nanfeng Township, mulberry, citrus, bamboo (for bamboo-shoot), chestnut etc. were evaluated. The increased benefits compared to the present values were estimated by multiple objective programming and indicative optimal plans determined which provided a basis for discussing models to be tested with local communities. Unfortunately, the extension sector cannot always replicate such an approach as it often concentrates on production technology alone and ignores the wider range of complementary technology and services.

3.5 Government Support and Socio-Economic Coordination

Recently, the Chinese Government has begun to give more attention to integrated agricultural development and ecological agriculture. During 1989, a national conference on integrated agricultural development held in Beijing concluded that, whilst the present resources must be well utilized to enhance the yield per unit area, greater attention must be paid to integrated development and utilization of new agricultural resources to enhance the integration of production. It was emphasized that investment must increase; that favourable policies are required; that there should be good scientific management; and that different agricultural sectors must be coordinated. The conference selected several regions and projects as key test sites, with a priority to hilly and dry areas.

The policy of individual household responsibility, effective since 1977, has been greatly welcomed by farmers and has contributed to restoring and enhancing the agricultural production to a major extent. When ecological agriculture has been practised, consolidation of fields has proved useful. Meanwhile, support services (input supply, processing and marketing) have been strengthened. Cooperation among farmers has been promoted.

4. Reorientation of Education, Research and Extension toward a Systems Approach

During the past forty years, ZAU contributed to the introduction of systems approaches. Graduate students in farming systems were first accepted by ZAU in the early 1950s. During the early 1980s, they organized a multi-disciplinary academic and technical research team with more than thirty professors and lecturers of nine disciplines (including agronomy, animal science, agricultural economics, rural sociology, ecology) which manages several outreach sites for integrated development and research and advises M.Sc. students. An Agro-Ecology Institute (AEI) was established in January 1989, taking agricultural ecology and farming systems as the main themes. In 1990, the

State Council authorized an agricultural ecology Ph.D. programme at ZAU, which is the first such authorized multi-disciplinary programme among all the agricultural universities.

FAO played an important role in facilitating this process. ZAU was the national coordination institution for Project RAS/81/044 "Education and Training on Integrated Farming Systems in Asia"; and in so doing, learned more about integrated farming systems from FAO staff and experts in Southeast Asian countries. It was demonstrated that the systems approach incorporating integrated farming systems, ecology, economics, systems engineering and decision-making, plays an important role in the process of rural development.

In connection with the proposed project RAS/86/040 concept, two new rural out-reach sites in the rainfed area were established in the Zhuma Township, Jinhua City and Sangqiao Township, Deqing County. Supported by the Governor of the Zhejiang Province and by leaders at various levels, strong technical and administrative organizations were established. Training included farmer training and a six-week trainers course for forty SMS from extension services.

5. Suggestions about Future Work

New mechanisms are required that integrate extension, applied research and training with a systems view and methodology. Under the system of people's communes, governmental organization and the effective decision-maker concerning agricultural production, ie. communes, acted together and were defined entirely by administrative orders. However, the old extension system was not appropriate after the introduction of the individual household responsibility system. Alternatively, in line with the proposed project RAS/86/040, decentralized experimental stations should integrate demonstration and applied research whilst being the centre of extension and training. Such stations should be established at the township level and be linked in a network.

In directing agricultural production, government agencies have the duty to lead and organize integrated agricultural development. Presently, closely related components within farming systems are handled separately by different agencies, each having an independent budget for its own work. A coordinating agency is needed with the authority to coordinate various agencies within agriculture through plans and projects, to centralize capital investment decisions according to the integrated agricultural plan. Such a reform is being tested in the Jinhua Municipality and Deqing County, Zhejiang Province. The provincial and central government should also be strengthened similarly. Presently, an office for integrated development at the provincial levels has been established for general coordination but has had limited impact so far.

INSTITUTIONALIZATION OF A FARMING SYSTEMS APPROACH TO DEVELOPMENT

The Institutionalization of Farming Systems Development (FSD) is a major challenge. One major priority area for intervention is extension. Not only must formal integration be promoted, but the informal multi-disciplinary methods of FSD are required in China to analyze current problems and devise local solutions. Standardized methodologies including rapid rural appraisal (such as ZAU is now teaching) should be disseminated widely, which requires a substantial expansion in training capacity. Unfortunately, the standard for assessment and promotion in extension is mono-disciplinary, which results in little systems orientation of extension workers, and is one of the major obstacles that retards the application of work like FSD.

Agricultural technicians and farmers need training. Farmers with knowledge of improved technology are the most powerful force for the dissemination of agricultural technology. Therefore training farmers is very productive. Meanwhile, agricultural technicians and officers at the basic level need refresher training especially to replace the "reductionism" methods learned in earlier times. Those trainees from the ZAU/AEI programme indicated that they were able to foresee problems more fully and had more tools of analysis, problem solution and technology dissemination after six weeks of professional training. The expansion of these training programmes would appear to be a priority for the effective institutionalization and application of ecological agriculture and the farming systems development approach.

FARMING SYSTEMS APPROACH
- KENYA EXPERIENCE -

by

G.G. Mwangi[1]

1. OVERVIEW

The Farming Systems Approach provides an alternative pathway and communication linkage for technological advance and agricultural development, without requiring major advancements in general education or major changes in the institutional orientation of national research systems. It complements the scientific concepts and procedures applied on experiment stations by providing a multi-disciplinary approach to understanding the complexity of the farmer's environment, farming systems and decision process.

It is clear that in some countries food deficits exist because of slow production growth, while in others food problems exist because of inadequate distribution resulting from poor roads and transportation facilities and other inadequacies of infrastructure. In yet other situations, food problems exist because the poor, lacking opportunities for employment, also lack purchasing power. Generally the impediments to increased production and increased purchasing power are closely intertwined.

It is to be hoped that the Farming Systems Approach can contribute at this as well as at the farm level:

a. To develop a strategy addressed to issues that bear on the pivotal role of food and agriculture in the development process.

b. To formulate an appropriate technology policy, emphasizing the essential function of technological change in accelerating agricultural growth.

c. To aim at alleviation of poverty by emphasizing concern on food consumption and nutritional status of low-income people and the critical importance of income generation for improving their status.

---

[1] Assistant Director Agricultural Engineering, Ministry of Reclamation and Development of Arid, Semi-arid Areas and Wastelands, Kenya

d. To look at the effects of food aid from the developed to developing countries. Food aid should only be given under extreme food shortage circumstances.

At the farm level, FSA should elucidate and seek to improve the farmer's own strategy, which typically is one of adopting technologies sequentially, based on availability, technical viability in the field, economic profitability and risk considerations and the resource endowment fit of the technology concerned within the farming system.

2. FSA IN KENYA

Agriculture employs over 70% of the country's population. This sector earns 30% of GNP and over 60% of the foreign exchange. Food processing, beverages and textiles, which represent over 75% of our industrial output, are processed from agricultural raw materials. It is evident therefore that poor performance of the agricultural sector has serious and immediate repercussions on the Kenyan economy.

The current estimated population of 22 million is expected to increase to 35 million by the year 2000. It is therefore necessary that food, income and employment be increased at correspondingly high rates. According to the Sessional Paper No. 1 of 1986, agriculture is projected to grow at 5.3% and provide the majority of the 6 million jobs that will be needed by the turn of the century as well as ensure food security.

The concept of the farming system approach to agricultural development was introduced in Kenya in 1975. Although the concept was appreciated, very little was done up to 1983, after which several meetings and workshops have been held in the country to expose administrators, policy makers and planners to FSA concepts. Different fora have been utilized to train research and extension staff in the concepts and procedures of farming systems.

Three distinct areas have been identified in FSA in Kenya. These areas are research, extension and training. From 1989 efforts have been made by the various institutions involved to harmonize terminologies and methodologies. From a recently held workshop, it was clear that the objectives of FSA are known and accepted, and that therefore the task ahead is the application of FSA, to generate and diffuse relevant technologies/recommendations for specific groups of farmers focusing on identified priority problems and constraints.

The objective in Kenya is to apply FSA to improve the productivity and sustainability of existing farming systems under different agro-ecological zones and socio-economic conditions.

For the success of this approach the following measures are being taken:

a. The farmer's needs, objectives and priorities are given due consideration in the formulation of both regional and national agricultural development strategies.
b. All participants (researchers, extension staff, trainers, farmers, policy makers, etc.) are jointly involved in the process of technology development.
c. Designing, adapting and evaluating appropriate production technologies to satisfy farmers' and societal needs.

Kenya recognizes that FSA is not a discipline in itself but is only an approach. This fact is taken into consideration in programme planning and training. The approach also recognizes that both technical and non-technical issues are equally important in agricultural development, especially in their influence on the production and productivity of the small-scale farmers who are the rural majority in this country.

There are very many institutions working with farmers in this country. Some of these are directly involved (eg. Ministries of Agriculture, Livestock Development, Research, Science and Technology) while others are facilitating institutions (eg. the Treasury, Ministry of Planning and National Development, Donor Agencies). However, the following have been identified as key institutions involved or likely to be involved in making FSA a success:

a. Ministry of Agriculture
b. Ministry of Livestock Development
c. Ministry of Research, Science and Technology
d. Ministry of Reclamation and Development of Arid, Semi-Arid Lands and Wastelands
e. Ministry of Environment and Natural Resources
f. Training institutions (universities, colleges, polytechnics, institutes, etc.)
g. Regional Development Authorities
h. Non-Governmental Organizations (NGOs)
i. International Agricultural Research Centres
j. Private Sector
k. Donor Agencies (playing a facilitating role).

In addition, there are ministries concerned with one component of infrastructure (eg. water, roads), as well as those concerned with social services, and the machinery of local government.

There is urgent need to strengthen the linkages between all these institutions and to coordinate their activities for the successful implementation of FSA. It is also necessary to have simultaneous interaction and linkages between farmers, extensionists, researchers, and trainers) in all farming systems activities.

To facilitate the strengthening of the linkages the GoK Task Force on FSA recommended that various committees be set up as follows:

a. District Farming System Teams (DFST)
b. Regional Research Centre Advisory Committee (RRCAC)
c. National Farming Systems Coordinating Committee (NFSCC).

The NFSCC is now a functioning body and the others have been established on a pilot basis in selected regions.

3. THE ROLE IN FSA OF THE MINISTRY OF RECLAMATION AND DEVELOPMENT OF ARID, SEMI-ARID LANDS AND WASTELANDS

FSA comes at a time when the GoK has created a new Ministry to reclaim and develop the arid, semi-arid areas and wastelands of this country. It is estimated that about 80% of Kenya's land is arid or semi-arid. Roughly 30% of the country's population and more than 50% of its livestock live in this area.

More precisely, an estimated 8.2 million people lived in the 22 ASAL (Arid and Semi-arid Lands) districts in 1989, including both pastoralists and farming communities. However, a high proportion of these areas are either poorly utilized or are in serious decline owing to poor use, resulting in degradation and destruction of the natural environment. The ASAL area is in urgent and serious need of reclamation, restoration and protection for human settlement, increased productivity and the provision of a healthy environment.

The ASAL population of livestock, according to the 1989 figures from the Ministry of Planning and National Development, is as follows:

	heads, in million	% of country's total
Beef Cattle	5.80	64
Dairy Cattle	0.72	24
Sheep	4.10	64
Goats	7.30	85
Camels	0.95	100

These figures clearly demonstrate the importance of livestock not only in ASAL but for the economy on the whole. Pastoralism predominates particularly in the drier areas with 200-500 mm. rainfall and less.

Generally, the ASAL climate is hot as well as dry. The rains are unreliable and unpredictable, hence one cannot simply consider annual rainfall as the major factor in ASAL development. Evapotranspiration is more than twice annual rainfall. From the seven agro-ecological zones (AEZs) recognized in Kenya, four zones (IV-VII) based on moisture availability describe the ASAL:

AEZ	% of country area	Average Annual rainfall (mm)
IV Semi-Humid	5	700-800
V Semi-Arid	15	550-700
VI Arid	22	300-550
VII Very Arid	46	200-300

Owing to the unfavourable natural environment and harsh climatic conditions, and considering the prevailing cultural and socio-economic situation of ASAL communities, water for both humans and livestock plays a vital role within ASAL development. Subsistence pastoralism in particular puts emphasis on milk production rather than on meat production, and hence requires large herds and well-distributed water points.

Traditional pastoralism in ASAL has been found to be less risky, more economical and less taxing on the environment than traditional dry land farming. No wonder then that the ASAL economy has evolved mainly towards a livestock economy with some cropping conducted during good rain years and in selected sites where water accumulates. However, the inter-play between agriculture and livestock will need to find a new balance as people's traditions, habits and perceptions change. It will be essential to institute close intersectoral and inter-agency collaboration in order to avoid duplication and undue competition and, hence, wastefulness.

The new Ministry clearly has a special role in seeking to achieve the necessary coordination. This role the Ministry is fulfilling by:

- publicising ASAL needs;
- attracting donor support for ASAL development in specific districts;
- encouraging and coordinating the input of other Ministries into ASAL development.

In this context, the Ministry is among those who are now promoting multi-sectoral and participatory planning and development, as implied by the adoption of FSA.

It is hoped that FSA will ensure that ASAL development is undertaken in harmony with the environmental considerations to which the GoK is fully committed. The new approach will strive to develop a closer relationship with already established Training Visit (T&V) systems, with a view to hastening reclamation and restoration of ASAL areas for settlement and increased productivity in a healthy environment.

The need for economic development in ASAL areas is heightened by the fact that available resources in high potential areas have sharply declined and become strained, necessitating that lower potential areas contribute more fully to the sustenance of our increasing population. The relationship between crops and livestock is of key importance in ASAL development in order to maintain soil organic matter for reasons both of soil fertility and moisture conservation.

4. CONCLUSION

For sustainability, a development strategy is needed that is responsive to specific community needs (pastoralism, women, children, small scale farmers, etc) as perceived by the people themselves. The Strategy must encourage the people themselves. The Strategy must encourage the people to participate in their own development so that in the end donor agencies will be mere facilitators of development rather than developers. Through the institutionalization of farming system development the farming communities must be facilitated to acquire simple and functional technologies that enable them to manage and maintain their own facilities once developed. As awareness and resource development begin to flourish, outlets must be found for agricultural produce and other products that are surplus to the communities' needs in order that income may accrue for investment in other social and material needs.

For the success of the new approach it is essential to institute close inter-sectoral and inter-agency collaboration in order to avoid duplication and undue competition and hence wastefulness.

REFERENCES

1. Cereal Technology Development - West African Semi-Arid Tropics (A Farming Systems Perspective by Purdue University).

2. Government of Kenya ASAL Strategy and Environmental Action Plan.

3. Government of Kenya Sessional Paper No. 1 (1986).

VII

SUMMARY AND CONCLUSIONS

SUMMARY AND CONCLUSIONS

by

M. Hall[1]

Despite the fact that over two decades have passed since the introduction of farming systems approaches, most research still neglects the socio-economic aspects of technology generation and application. Extension services may be more knowledgeable about farmers' circumstances, but continue to promote the same sort of single commodity/package of practices advice that ignores enterprise interactions and resource limitations. There is an almost complete absence of the application of a systems perspective to other agricultural support services, to the design of area projects and to policy analysis. Most agricultural development workers agree that inter-disciplinarity is a good thing but continue to work in a mono-disciplinary fashion. All of them support the concept of participatory development but few people attempt to work with farmers, let alone learn from them.

The institutionalization of farming systems approaches has obviously not made much progress and efforts to date have been feeble. The Technical Discussions held in Rome, 15-17 October 1991, focused upon the identification of the major constraints responsible for this situation, as well as upon specifying short-term actions and longer-term programmes necessary to overcome them. This section of the document attempts to summarize the conclusions of the discussions with reference to this focus.

<u>Confusion of Terms:</u>   The subject of farming systems has generated a mass of different labels attached to variations of the basic approach. Most systems approaches have evolved from earlier ones, so there are many similarities between them and confusion concerning the definition of different approaches is to some extent inevitable. The confusion has, however, tended to create a negative feeling

---

[1] Research Associate, Harvard Institute for International Development

towards the subject on the part of some development practitioners. This attitude is understandable but unnecessary, since it is relatively simple to distinguish between the different variations of what is a basic theme.

All variations of the farming systems approach consist of an analytical framework or structured context, which gives a holistic perspective of any give farm situation. This facilitates the understanding of the relationships between the activities of the farm family and the forces at work in the surrounding environment. It also develops an appreciation of main priorities of the farm family and the types of decisions that they must make in the face of uncertainty and limited resources. In addition to this structured perspective, the approach adheres to certain basic principles which can be summarized as follows:

- Family centred
- Problem solving
- Participation
- Holistic analysis
- Inter-disciplinarity
- Respect for local knowledge

Farming Systems Development (FSD) is an attempt by FAO to synthesize the above approaches and the methodologies which they incorporate, in order to use the understanding of farm-family systems to improve a wide range of agricultural support services, area development activities and certain aspects of agricultural policy analysis. It is important for proponents of FSD to make it clear that theirs is an evolutionary, not a revolutionary approach, that seeks to build upon the strong points of previous approaches. FSD simply seeks to extend the scope of analysis and application of the farming systems perspective without in any way seeking to denigrate existing approaches.

Specialist Education and Scientific Method: Among the difficulties facing the successful introduction of a farming systems perspective, perhaps the most important are the attitudes and convictions of the personnel involved. Many senior agricultural professionals have been trained as specialists. Most have studied a mixture of physical and biological sciences and have been taught the classical reductionist scientific method of isolating a particular variable while keeping all others unchanged. Even economists are taught to analyze in this way (the phrase ceteris paribus appears frequently in economic documents).

This lack of training concerning complex interactions between multiple variables (holistic analysis) is exacerbated by the cultural gap which engenders a tendency to discount completely the priorities, knowledge and experience of smallholder farmers. Since farmers are forced to think in holistic terms, any discussion with educated specialists is almost a "dialogue of the deaf". Unfortunately, very few specialists have been taught to interact with farmers, or to understand the factors that cause them to do what they do.

As a result of this situation, many agricultural professionals have a continued faith in orthodox technology transfer approaches, and are not motivated to work in an interdisciplinary fashion. It is not easy to change the work habits of a lifetime, especially when reward systems are geared towards successful specialists.

Training and Educational Initiatives: The best way to improve the problem of work methods and professional attitudes is to introduce courses in farm-household systems into the curriculum of relevant colleges and university faculties. In the short-run, in-service training activities could influence key individuals to experiment with the approach in sections and departments that they control. However, there is a scarcity of suitably trained and experienced manpower needed to teach a farming systems perspective. Few specialists are tempted to become involved in farming systems teaching and research. It does not attract much academic prestige, so incentives for orthodox teachers to change are minimal. The few who are motivated find it difficult to acquire the necessary reorientation while sustaining a full teaching load. In addition, training materials are extremely limited and venues for publishing farming systems research fairly restricted.

Progress in institutionalizing the farming systems approach, obviously depends to a great extent upon successful training programmes. FAO and certain international research institutions have prepared training materials, but their scope is still limited. In order to overcome this shortage, networking of practitioners could be encouraged so that experiences and technical information can be exchanged and further training materials prepared. International organizations, and in particular FAO, should allocate resources for helping colleges and universities to design suitable curricula. Perhaps the action that currently constitutes the major constraint to introducing the approach, however, is the shortage of trainers - this should be addressed as a matter of priority.

Organizational Measures and Work Programmes: The integrated nature of rural life is seldom reflected within development institutions. These generally focus on a single sector (eg.agriculture), sub-sector (eg.livestock), or upon a particular aspect of development (eg.credit). Institutions dealing with agricultural development are commonly dispersed between several different ministries and parastatal agencies. Most are staffed by specialists trained to work on single components or sub-components of the agricultural development process. Work programmes, reporting systems and budget funding are all vertically arranged making it very difficult to function with any degree of inter-disciplinarity.

In the case of individual institutions, experience with the introduction of previous development perspectives indicates that complete restructuring will usually be unnecessary. A series of "soft" adjustments, such as modifications to staffing (including re-orientation courses) working methods, reporting procedures and budget allocations should suffice. The adjustments will almost certainly involve the secondment of specialists between ministries. They will also necessitate the recruitment of specialists in new disciplines, such as sociology and economics, into traditionally technical and scientific strongholds.

This series of modifications can only be really successful, however, where radical action is taken by government to decentralize decision-making and budgetary allocation. The reform process should permeate regular programme structures and not simply be confined to short-term projects. Decentralization also implies greater power being allocated to the community level, with more responsibility, authority and resources being devolved to farmers.

FAO and other relevant donors and technical assistance agencies should respond positively to requests from governments seeking to introduce a farming systems approach through institutional changes. As well as training materials, visits from experienced practitioners will be necessary in order to train a requisite core of national trainers. They will also need time to study current staffing patterns, work organization and methods with a view to recommending the necessary changes.

New Partners in Development: Agricultural development has largely been viewed in the context of a partnership between government organisations (GOs) and farmers organisations. The partnership is more complex in reality as, apart from the vital role of private sector commerce, non-governmental organizations (NGOs) have also demonstrated the need for their inclusion on equal terms.

The contrasting strengths, weaknesses and areas of interest of the two types of organization (GOs and NGOs) are complementary in many respects. They suggest a potential for closer collaboration and a further way to institutionalize the farming systems approach. NGO ideas and initiatives, plus their skills in relation to the analysis of farm-household situations and organization of communal participation, could be combined advantageously with governmental budgetary and logistical support, plus strength in technical and scientific areas. The breadth of GO influence could allow the scaling-up of successful, but limited, NGO approaches including those that have adopted a measure of farming systems thinking.

Lack of Success Stories: In addition to re-orientation of individual attitudes, adjustments to institutional processes, and forging a partnership with NGOs, more efforts need to be concentrated on recording and publicizing successful introductions of systems approaches. A great deal of effort has gone into defining concepts and theorizing about the applicability of farming systems approach to different aspects of agricultural development, but little progress has been made to document success stories. FAO should include this aspect of institutionalization in its regular work programme by generating such documentation, plus suitable audio-visual material, for inclusion in orientation seminars and workshops.

Absence of Detailed Methodologies: Although it is frequently stated that a farming systems perspective should underlay the work involved in most aspects of agricultural development, there is an almost complete absence of documented methodologies for FS application in all fields except technology development (FSR). It would seem that all the necessary ingredients are in place to develop and document these methodologies - analytical framework, underlying principles and tools to apply them. What remains is the preparation of a series of manuals that can be used by trainers and practitioners.

These efforts should include the development of specific work procedures and methodologies to integrate the process of technology generation with its dissemination, adaptation and application. This involves the introduction of a systems perspective into agricultural support services and programming activities, largely by adapting and strengthening existing methodologies. Manuals for the preparation of area development programmes and for applying a farming systems perspective to agricultural policy analysis are also needed. One means of gaining momentum for this task would be to organize a second set of Technical Discussions focused specifically on this area.

APPENDIX

APPENDIX

LIST OF PARTICIPANTS

People's Republic of China	Prof. Zhao Qian Wang Department of Agronomy Agricultural Institute Zhejiang Agricultural University Hangzhou Zhejiang 310029
France	Mr. Philippe Jouve Scientific and Pedagogic Director CNEARC 2247 av. du Val de Montferrand BP 5098 34033 Montepellier Cedex 1
Germany	Prof. Werner Doppler University of Hohenheim Windhalmweg 10 Postfach 700 562 D-7000 Stuttgart 78
Indonesia	Mr. Iksan Semaoen INRES Project Brawijaya University Kotak Pos 176 Malang 65101
Kenya	Mr. G.G. Mwangi Ministry of Reclamation & Development of Arid, Semi-arid Areas & Wasteland (ASAL) Nairobi
Philippines	Mr. William Dar Bureau of Agricultural Research Department of Agriculture Ati Bldg. Elliptical Road Diliman, Quezon City

Philillines, contd   Mr. Tung LY
Farm and Resource Management Institute
Visayas State College of Agriculture
Visca, Leyte 6521-A

Mr. V.P. Singh
Agroecology Unit, APPA Division
International Rice Institute
PO Box 933
1099 Manila

Sri Lanka   Mr. Nimal Ranaweera
President of Asian Farming Systems Association
Division of Agricultural Economics & Projects
Department of Agriculture
 O Box 7
Peradeniya

Sweden   Mr. Ingemar Croon
International Rural Development Centre
Swedish University of Agricultural Sciences
(SUAS), PO Box 7005
S-750 07 Uppsala

Thailand   Mr. Basilio N. de los Reyes
Regional Farm Management Economist
FAO's Regional Office for Asia and the Pacific
Maliwan Mansion
Phra Atit Road
Bangkok 10200

United Kingdom   Mr. Anthony Bebbington
Centre for Latin American Studies
Cambridge University
History Faculty Building, West Rd.
Cambridge CB3 9EF

Mr. Stephen Biggs
University of East Anglia
Norwich NR4 7TJ

Mr. Andrew Shepherd
The University of Birmingham
Edgbaston
Birmingham B15 2TT

United Kingdom, contd Mr. John Farrington
Overseas Development Institute
Regent's College, Inner Circle
Regent's Park
London NW1 4NS

Mr. David Gibbon
School of Develoment Studies
Overseas Development Group
University of East Anglia
Norwich NR4 7TJ

FAO Mr. Karl H. Friedrich, Senior Officer (Farming Systems), AGSP
Mr. John Dixon, Agricultural Production Economist, AGSP
Mr. Marc Bral, Chief, AGSP
Ms. Ursula Löwe, Associate Professional Officer, AGSP
Mr. Martin Bostroem, Associate Professional Officer, AGSP
Ms. Maria Richardsdotter, Associate Professional Officer, AGSP
Mr. Malcolm Hall, Consultant
Mr. David Pratt, Consultant